Advance Praise for *Renewal*

Rediscovering joy through nature is of critical importance within the climate movement; the bonds that hold movements together aren't just strategic, they build our compassion. This book can be a guide on that journey.

— May Boeve, executive director, 350.org

Renewal is a journey that takes us home.... Edwards' personal love affair with this world and his eclectic relationship with sages from every time and culture make this journey delightful, nourishing, and worth every page.

— Sandy Wiggins, co-founder and principal, Consilience, LLC; Director, Business Alliance for Local Living Economies (BALLE)

This inspiring mix of inner work and outer action reminds us we still have a chance to save the planet (and ourselves), if we turn to nature as our guide.

— Mary Reynolds Thompson, author, *Embrace Your Inner Wild* and *Reclaiming the Wild Soul*

Andrés Edwards' important new book walks us through the steps of forging a powerful emotional bond with the rest of nature...

— Linda Buzzell, co-editor, *Ecotherapy: Healing with Nature in Mind*

A key subject, explored with clarity.

— Jeremy Narby, author, *Intelligence in Nature*

Renewal contains many guiding ideas and suggestions for making the connection [with nature] a real part of our lives. After you read it, get out there and build your own lifetime of experiences.

— Carl Safina, author, *Beyond Words: What Animals Think and Feel*

Saving nature is not just an altruistic act. In saving nature we also save ourselves since our innate affiliation with it supports our well-being. Nature provides services that we can hardly replicate and can be the inspiration for the things we design. Andrés Edwards has articulated the message that our lives are tied economically, intellectually, and spiritually to nature.

— Bill Browning, Partner, Terrapin Bright Green

This book is a pioneering exploration of an epochal (and bitterly necessary) shift in our attitude toward nature. Edwards sees nature not as an external resource to make human life safe but as part of the paradigm of reciprocity that makes our existence possible and allows reality to flourish.

— Andreas Weber, author, *Biology of Wonder, Matter & Desire,* and *Enlivenment*

This book includes remarkable stories and neuroscience discoveries that inspire us and call on us to connect to nature and live more fulfilling lives.

— Thupten Jinpa, author, *A Fearless Heart*

Robinson Jeffers wrote: "We must uncenter our minds from ourselves; we must unhumanize our views a little, and become confident as the rock and ocean that we were made from." Andrés Edwards confidently and elegantly guides us through how to unhumanize, and then rehumanize, ourselves.

— Wallace J. Nichols, PhD, author, *Blue Mind*

In these times when less than half of Americans say they participate in outdoor recreation, and mental health problems have become epidemic in part due to people spending too much time watching electronic screens, Andrés' book will help people return to their roots and discover the healing and inspiring powers of nature.

— James A. Swan, PhD, author, *Nature as Teacher and Healer*

Using personal anecdotes and scientific evidence, *Renewal* illuminates the different ways we can emotionally and intellectually connect with nature as well as practical ways we can deepen this relationship to promote both our own flourishing, and that of the natural world around us.

— Craig L. Anderson, PhD, University of California, San Francisco

Renewal

Renewal

How Nature Awakens Our Creativity, Compassion, and Joy

Andrés R. Edwards

Foreword by Marc Bekoff

To Michael & Paula,
May your creativity
Keep blossoming in
nature!

[illegible]
1/18/2024

new society PUBLISHERS

Cover design by Diane McIntosh. Photo © iStock 486974672.
Text images: p. xvi © magann; p. 134 © Peter Hermes Furian;
tree silhouettes © gorralit / Adobe Stock.

Printed in Canada. First printing April 2019.

To order directly from the publishers, please call toll-free
(North America) 1-800-567-6772,
or order online at www.newsociety.com

Any other inquiries can be directed by mail to:
New Society Publishers
P.O. Box 189, Gabriola Island, BC V0R 1X0, Canada
(250) 247-9737

Funded by the Government of Canada | Financé par le gouvernement du Canada | Canada

LIBRARY AND ARCHIVES CANADA CATALOGUING IN PUBLICATION

Edwards, Andrés R., 1959–, author
Renewal : how nature awakens our creativity, compassion, and joy / Andrés R. Edwards ; foreword by Marc Bekoff.

Includes bibliographical references and index.
Issued in print and electronic formats.
ISBN 978-0-86571-880-7 (softcover) ISBN 978-1-55092-673-6 (PDF)
ISBN 978-1-77142-268-0 (EPUB)

1. Human ecology—Psychological aspects. 2. Nature—Psychological aspects. 3. Human beings—Effect of environment on—Psychological aspects. 4. Well-being—Psychological aspects. 5. Nature, Healing power of. 6. Nature—Effect of human beings on. 7. Human ecology. 8. Nature. 9. Well-being. I. Bekoff, Marc, writer of foreword II. Title.

BF353.5.N37E39 2019 155.9'1 C2018-905823-4
C2018-905825-0

New Society Publishers' mission is to publish books that contribute in fundamental ways to building an ecologically sustainable and just society, and to do so with the least possible impact on the environment, in a manner that models this vision.

To future generations of practical visionaries
who embrace a reciprocal relationship with nature
for the benefit of all life

Contents

Acknowledgments

The inspiration for this book came to me several years ago during an afternoon bike ride. Since that day, I've been blessed with the insights of many who have helped to shape the book. I am thankful to the following colleagues and organizations for their valuable insights into our relationship to the natural world: Elizabeth Thompson, Megan Ahern and J.P. Harpignies formerly of The Buckminster Fuller Institute; Center for Compassion and Altruism Research and Education; Greater Good Science Center; Stacy Carlsen from the Marin County Livestock & Wildlife Protection Program; M. Ananda Kumar and Smita Prabhakar from Nature Conservation Foundation; Peter Sherman from Prescott College; and Camilla Fox from Project Coyote.

I would like to thank the following friends and colleagues who, through our numerous lively discussions over the years, have helped me to take a "deep dive" into the perspectives expressed in this book: Robert Apte, Laurent Boucher, Phyllis Mufson, Jim Newell, Greg Newth, Jeff Reynolds, Lia Rudnick, Mark Samolis, Nadine Ulloa, Nils Warnock, Don Weeden and Mark Woodrow.

I am indebted to the following for their photographs, which have been helpful in illustrating the stories described herein: Dave Alan, Bill Browning, s-eyerkaufer, fotogaby, ithinksky, P. Jeganathan, Cathy Keifer, Ian McDonnell, Andrea Pavanello and Riorita.

I am grateful for the unending dedication of my editor Diane Killou, who reviewed my drafts from the very beginning and gave me encouragement throughout the writing process. Thank you to Rand Selig and James Swan, who kept an eye on the "big picture" and helped me discern what's essential and what's not. I'm indebted to the New Society Publishers team including: Ingrid Witvoet, E.J. Hurst,

Judith Plant, Sue Custance, Greg Green, Diane McIntosh, Sara Reeves, Julie Raddysh and Jean Wyenberg, who have been instrumental in the birthing of this and my previous works over the last decade.

My life's journey has been enriched by Kathleen Walsh and by my children, Naomi, Easton and Rylan. Thank you for being my teachers and embodying brilliant possibilities for future generations.

Foreword

Why care about nature? This question delves deep into our values. The reasons to care about nature range from the practical to the philosophical. First, nature keeps us alive; second, nature enriches our lives; and third, as the dominant species on Earth, we have a moral imperative to care for the well-being of all humans, all non-human species and the environment.

Nature provides us with the air we breathe, the soil to grow food, the water we drink and the wood, earth, concrete and stones to build our shelters. Our economic industries such as fisheries, forestry, mining, energy, construction, agriculture and tourism rely on natural resources to flourish. Our survival depends on nature's resources and the work of millions of diverse species—ranging from pollinators to photosynthetic plants to the nutrients in the soil—that form a vital part of our vibrant and magnificent planet.

Nature enriches our lives through her lessons, her beauty and her healing. Being in nature can reduce our stress and blood pressure and make us happier and more creative. Witnessing spring wildflowers, trees, insects, birds and mammals can enhance our lives and spark our curiosity, appreciation, altruism and compassion. With 3.8 billion years of experience, nature is a teacher and mentor showing us how to design technologies that adapt to the natural world without degrading it.

Taking care of the Earth is a moral imperative. As the species with the greatest impact on the planet, we can make daily choices to support life on Earth. As primatologist Jane Goodall points out, "Every single individual matters, every single individual makes some kind of impact on the planet every single day. And we have a choice as to what kind of difference we're going to make."[1] Being conscious of the choices we make each and every day is the first step in understanding the ripple effect that our decisions have on us, on other animals and on the planet's health. These choices range from the products we purchase to modes of transportation, the clothes we wear, the food we eat and the family and business decisions we make. Each choice can make an important difference in the world.

Our choices are based on our values. *Renewal* explores how we can nurture an ecocentric ethic, which encompasses a reciprocal relationship with nature where we use natural resources wisely and enhance the biodiversity of nonhuman species. Compassionate conservation, which focuses on the health of each individual, is a values-based approach to wildlife conservation. Taking life-affirming actions that benefit all species is at the core of an ecocentric ethic, which will serve as a moral compass as we navigate the balance between the needs of human and nonhuman species.

What is our role in nature? As the dominant and ever-growing species, do we embrace the role of exploiter or land steward? Are we anthropocentric or ecocentric? Are we working toward short-term immediate rewards or a long-term vision that benefits all species? The answers to these questions determine what our legacy will be.

When we align with nature's rhythms we can view our role as humans within the larger context of the tapestry of life. It's no longer just about us, but also about all living beings. The seasonal cycles, the growth of plants, the movement of clouds, rushing rivers and active insects all mark the constant motion of the world. Finding how we fit into this living web helps us feel more connected to all life.

When we connect to the natural world, we have a chance to discover who we are and how we can be of service to others. Whether we immerse ourselves in nature intellectually, emotionally, recreationally or spiritually, in the end we gain a new perspective on how we fit into the deeply interconnected world.

Renewal helps us to rediscover the bountiful gifts we can receive from nature when we open ourselves to our experience without expectations. By taking a deep dive into the intelligence of nature as exemplified by the numerous insect, plant and animal stories shared in this book, we learn that we have a vital role to play in maintaining healthy ecosystems.

As our natural habitats come under increased pressure from rapidly growing human populations around the world, Andrés Edwards' timely and important book reminds us that many of the solutions to our global problems lie in reacquainting ourselves with the natural world—what I call "rewilding our hearts." Understanding how animals build shelters and plants produce food and generate energy shows us how we can adapt to the global challenges we face. Learning about the behavior of other animals also gives us valuable insights that can help us nurture our compassion and humility toward one another and the natural world.

When we move beyond our intellect and feel the power of being in nature in our hearts and in our very souls, we plant the seeds for caring about and loving the natural world. The stories in *Renewal* encourage us to remember the value of a strong connection with nature and invite us to explore how we can deepen this bond in an age when many people—far too many people—are becoming alienated from nature and, in the process, becoming alienated from other humans and from themselves.

— Marc Bekoff
Boulder, Colorado
January 2019
Author of *Rewilding Our Hearts:*
Building Pathways of Compassion and Coexistence

INTRODUCTION

Forging an Emotional Bond with Nature

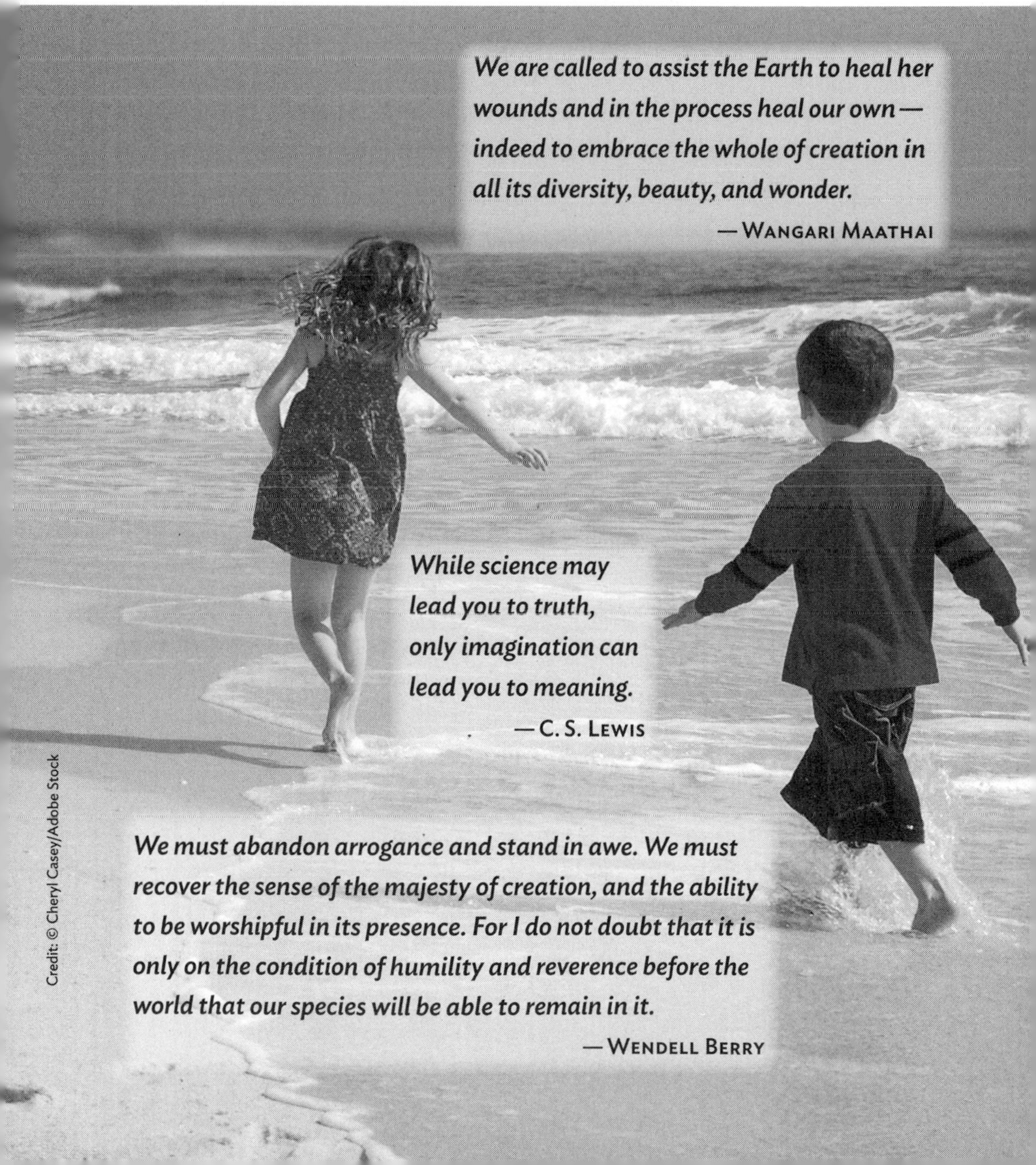

We are called to assist the Earth to heal her wounds and in the process heal our own — indeed to embrace the whole of creation in all its diversity, beauty, and wonder.

— Wangari Maathai

While science may lead you to truth, only imagination can lead you to meaning.

— C. S. Lewis

We must abandon arrogance and stand in awe. We must recover the sense of the majesty of creation, and the ability to be worshipful in its presence. For I do not doubt that it is only on the condition of humility and reverence before the world that our species will be able to remain in it.

— Wendell Berry

Credit: © Cheryl Casey/Adobe Stock

Evolutionary biologist Stephen Jay Gould declared, "We cannot win this battle to save species and environments without forging an emotional bond between ourselves and nature as well—for we will not fight to save what we do not love."[1] As one of the "younger" species inhabiting planet Earth, we humans have embarked on an epic journey to redefine our relationship with the natural world. Our journey begins with the cognitive science breakthroughs that are revealing the impact of nature on our behavior and emotions, and expands outward to encompass a compassionate way of coexisting with nonhuman species and the air, soil, water, minerals and ecological processes that support all life on the planet.

Since we have gradually forgotten the importance of nurturing our emotional bond with nature, we are in a new epoch of remembering. Native peoples such as the Salish from the Pacific Northwest embraced a state of mind where we use our hearts to live by and to help the power, beauty and magic of nature flourish. In more recent times, environmentalist Rachel Carson reminded us that "it is not half so important to *know* as to *feel*," emphasizing the importance of our emotional connection to nature rather than relying solely on our intellect.[2]

Biologist E. O. Wilson expands on our emotional connection to nature through the biophilia hypothesis, which describes our "innately emotional affiliation" to living organisms.[3] And marine biologist Wallace J. Nichols promotes our emotional bond with nature through neuroconservation, focusing on developing a conservation strategy rooted in our neurological responses to nature, especially water. As Nichols points out, "It's time to drop the old notions of separation between emotion and science.... Emotion is science."[4] All of these ecological visionaries show how we need to rekindle our feelings about nature and blend our scientific breakthroughs with our emotions.

Recent cognitive studies, aided by technologies such as the CAT scan and the fMRI, have shown numerous physical, behavioral and emotional benefits from being in nature. These include being healthier through reduced stress, blood pressure levels and risk of cancer; and being happier, more compassionate, grateful and creative. But more important than what we take from nature is what we give back.

An ecocentric ethic asks: "What is our responsibility as stewards to give back to the natural world?" One way of giving back is by embracing a compassionate way of living and developing restorative initiatives that help people, other species and the environment to thrive. This is a reciprocal relationship rooted in embracing our interdependence with nature and taking actions that enrich our connection with it. Douglas Christie reminds us that "our ecological commitments, if they are to reach mature and sustainable expression, need to be grounded in a sense of deep reciprocity with the living world."[5] This reciprocity beckons us to shift away from short-term objectives and quick fixes and instead adopt a long-term, resilient vision for the future—one in which we play an integral role and take responsibility for its fruitful outcome.

Renewing ourselves and nature also involves a biomimetic approach in which nature is our mentor and teacher. We are already using nature's 3.8 billion years of experience to learn how to generate abundant renewable energy, grow healthy food crops without depleting the soil and water table, provide safe drinking water, design efficient transportation systems and access to medicines and develop new ways to eliminate waste and pollution and stabilize the climate.

We have much of the knowledge needed to achieve these objectives. Now we need to streamline the social-political systems that act as barriers. We can do this by remembering ourselves as compassionate beings who care for one another and for the environment. Taking care of each other and nature begins by emulating nature's living systems so that we live in harmony with it. This approach is based on a model not of scarcity but of abundance. It involves recognizing that although we have an important role to play as a dominant species, we depend on nature for our survival. It's a relationship where we "give" and "take" so that everyone thrives.

Our relationship with nature also benefits when we practice the precautionary principle ("better safe than sorry"). When we consider our responsibility as stewards of the Earth with humility, we gain a broader perspective to make wiser decisions that affect all life on Earth. Many of the planet's global systems, such as the climate, are impacted by our actions. Following the precautionary principle in implementing a new technology, we take action only after ensuring a safe outcome.

By nurturing our innate curiosity and our affinity for nature we can renew our respect and admiration for the natural world. We are learning, for instance, about the remarkable ability of bees in designing their hexagonal-shaped honeycombs, crows in communicating dangers across generations, caribou herds using swarm intelligence and evading wolves with precise movements and trees that communicate with each other about impending droughts. These examples ignite our passion for nature's genius. This passion is a recipe for falling in love with and protecting nature. Witnessing nature's genius stimulates the creativity we need to devise ways to enhance rather than degrade the environment.

The altruism of nonhuman species inspires us to emulate their acts in our families and our communities. Brazilian ant species sacrifice themselves to protect their kin by sealing the colony's entrance and dying in the cold overnight temperatures; female bats share regurgitated blood to nourish other bats in need; and honey bees fatally rupture their abdomens after using their stinger to protect the hive. These altruistic acts illustrate how nature mirrors the best qualities in the human heart.

Nature can teach us how to live compassionate, creative and joyful lives. Our hearts grow as we remember the importance of loving nature. I hope the stories in this book inspire you to discover how you can make your life and nature thrive by nurturing a reciprocal, enduring relationship with the natural world.

1 Aligning with Nature

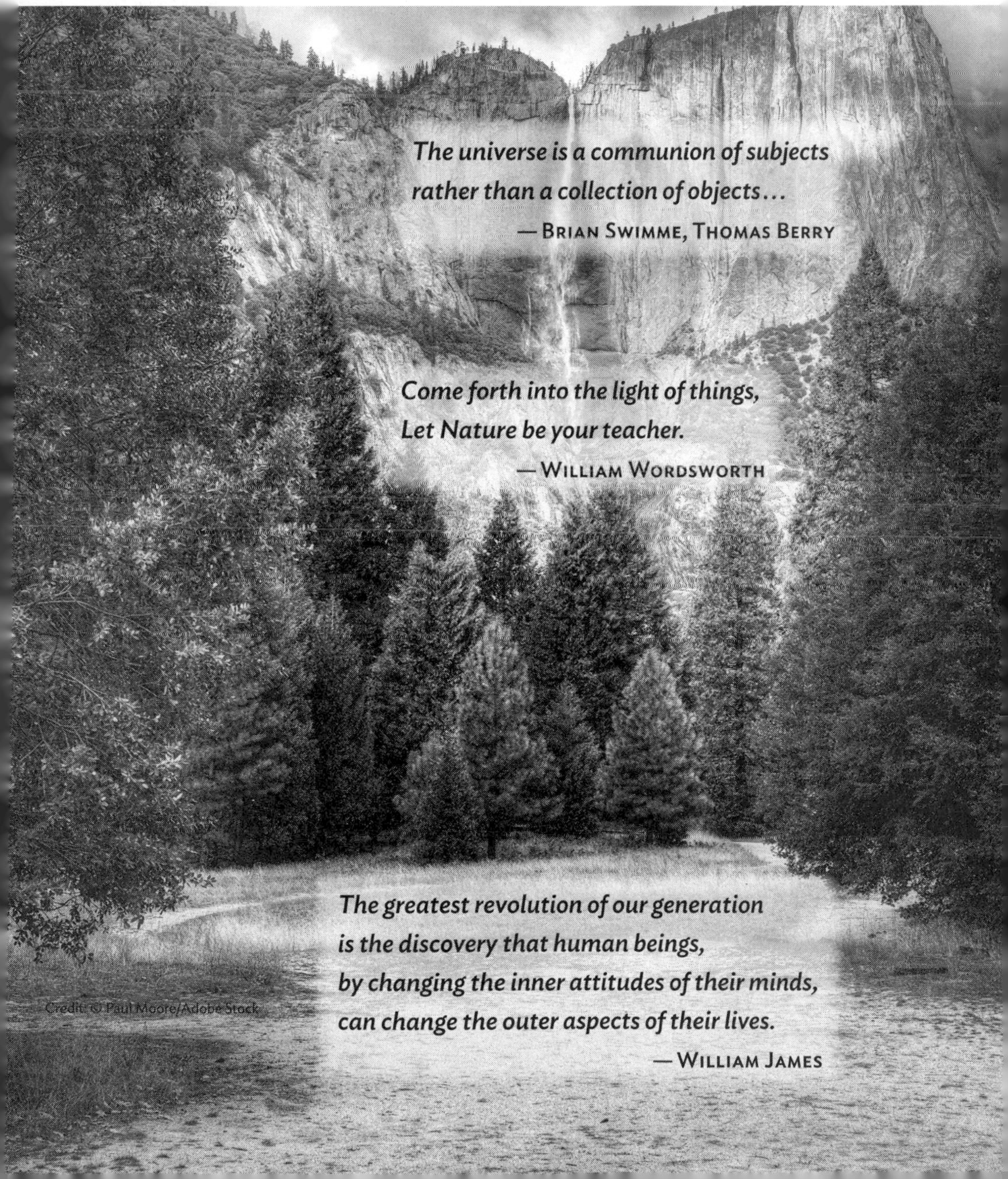

The universe is a communion of subjects rather than a collection of objects…
— Brian Swimme, Thomas Berry

Come forth into the light of things,
Let Nature be your teacher.
— William Wordsworth

The greatest revolution of our generation is the discovery that human beings, by changing the inner attitudes of their minds, can change the outer aspects of their lives.
— William James

Credit: © Paul Moore/Adobe Stock

BEING ALIGNED WITH NATURE CAN COME IN handy when one's life is on the line. I vividly remember a sunny summer afternoon when three of us decided to climb a nearby rock face. I was halfway up when there was a mix-up and I found myself high above the ground without a rope. My friend who was above me tried several times to throw a rope down to me so I could tie myself and climb up to him, but because he had no line of sight, every time he threw the rope it landed several feet away from me. It was starting to get dark and I had to make a decision. I had three options: have my friend keep trying, hoping that the rope would eventually reach me, stay put and wait for a rescue or climb unprotected toward the rope, which was about 20 feet away. I decided to climb.

I knew there was no room for error but I also knew it was not a particularly difficult climb for me. I just needed complete alignment with the rock. As I made my way toward the rope, time stood still and every move flowed seamlessly. I focused on breathing and being present, grounded and fully aware of my body's movements and my connection to the rock. Nothing else mattered. In this attuned state, I made my way toward the rope with ease and assurance. Time was suspended. It may have taken me one minute or five minutes. What was paramount was my alignment with the rock. Everything else around me—the fading sunlight, the breeze and the surrounding trees and mountains—disappeared into the background. When I reached the rope, I had a huge feeling of relief followed by a physical and emotional exhale that enveloped me.

Being aligned with nature conjures an image of integration, cooperation and flow, similar to a current meandering through a stream or a breeze blowing through a forest canopy. We probably can all recall a few instances in our lives when we've felt a sense of unity with nature and with life while experiencing a special moment in the natural world—perhaps a walk in a park with a child, discovering a bird's nest, seeing a butterfly's intricate wing pattern, noticing a spider's web in the sunlight or marveling at a starry night. These experiences awaken our sense or curiosity and reveal the mystery and beauty that surround us. However, in a world where more than half the population lives in urban areas and people spend 90 percent of their time indoors, this connection with nature is rare indeed. As our separation from the land and each other increases, the need to find viable ways to align with nature has become imperative. How can we realign with nature? The answers vary depending on our temperament, cultural background and interests.

Doorways to Nature

The intellectual door into aligning with nature is our innate curiosity about the natural world. We may be drawn to learning about how nature works. This scientific approach builds on previous investigations and its goal is a new discovery or an improvement in the understanding of a process or an event. As technology matures, scientific discoveries are happening at an accelerated pace but the first step is asking a question, being curious about why something happens, followed by a possible explanation, which is modified based on the findings. Being curious is an innate human quality that is especially present during childhood and for many of us it continues throughout our lives. Nature is an ideal place to ignite our curiosity.

The spiritual door into aligning with nature goes back to indigenous cultures, where humans are seen as part of the web of life. Human history is intertwined with the Earth's evolution and is seen in the creation stories of the First Peoples from all continents—Aboriginal, Native American, Amazonian, European and African tribes. Through their ceremonies and stories these cultures have found enduring ways to remain aligned with nature. As author Wilma Mankiller notes, "indigenous people have the benefit of being regularly reminded of their responsibilities to the land by stories and ceremonies. They remain close to the land, not only in the way they live, but in their hearts and in the way they view the world. Protecting the environment is not an intellectual exercise; it is a sacred duty."[1] The spiritual alignment with nature prevalent in indigenous cultures has an important connection to our collective responsibility to see ourselves as stewards of the land. As an integral part of the web of life we have the choice to take care of the land and the species that inhabit it.

The spiritual door into nature also beckons us to explore the meaning of the natural world in our lives. For some of us, nature acts as a spiritual sanctuary that provides us with clarity during challenging times. Witnessing the infinite expressions of nature—through torrential thunderstorms, ancient trees, saturated sunsets, resilient plants, curious animals, relentless insects, exquisite flowers and meandering streams—often fills us with wonder and humility.

Finally, there is the experiential door into nature—getting out there and being in the wilderness. The underlying motivation can range from education to a desire for personal growth and well-being to simply relaxing and having fun. In *Multiple Intelligences: New Horizons in Theory and Practice*, developmental psychologist Howard Gardner describes "naturalist intelligence" (one of several human intelligences), which focuses on our affinity with nature.[2]

I recall my earliest memories of fishing and camping with my family and marveling at the opportunity to be among the trees and rivers and lakes, away from an urban environment. In college I read the works of Thoreau, Emerson, Carson and Leopold among others, gaining a deeper connection to nature. Later I led groups into the backcountry and worked with youth in the outdoors, cementing my appreciation of and caring for nature. But the seed was an emotional bond that took root in my childhood.

Nature helps us get out of our heads and into our hearts. Experimenting with our senses in nature is a powerful way to increase our awareness of everything around us. When we close our eyes, for instance, we often hear more acutely and may feel the breeze on our face and sense the texture of the soil beneath our feet. Focusing on the smells of spring flowers and the taste of wild berries gives us renewed appreciation for nature's bounty. And sometimes when we sit in one place for an extended period of time and observe everything that's happening, we notice that remarkable stories are unfolding all around us. Perhaps it's the termites meandering toward their mound or the ever-changing patterns of the clouds or the fish that briefly jumps up above the pond's surface. These ongoing stories provide us with a front row seat on life as it unfolds every second of every day. What a gift! As Jack Kornfield said, "The present moment is all we have, and it becomes the doorway to true calm, your healing refuge."[3]

In our modern culture, experiencing nature includes a mixture of natural and artificial sounds. The soundscape of nature is made up of nonbiological sounds such as wind, rain and thunder, known as geophony; sounds from living organisms such as birds, dogs and dolphins, known as biophony; and sounds from humans such talking, music, cars and airplanes, known as anthropophony. All these sounds make up the living tapestry of many wilderness areas. I remember once spending many days in a remote desert in Utah immersed in a biophony without hearing or seeing another human for days, yet experiencing moments of anthropophony as passenger jets flew high above me. I've also enjoyed the biophony of remote wild areas where only bird sounds greeted me at the dawn of a new day.

Alonzo King's and Bernie Krause's *Biophony*, which integrates ballet with natural sounds, brings to life the richness of nature's soundscapes. Krause writes, "This is the tuning of the great animal orchestra—the inspiration for the ballet. It's an illumination of the acoustic harmony of the wild, the planet's deeply connected expression of natural sounds and rhythm. It is the reference for what we hear in today's remaining wild places, and it is likely that the origins of every rhythm and composition to which we dance come, at some point, from this collective voice. At one time there was no other acoustic inspiration."[4] Artistic performances such as *Biophony* illustrate the powerful lure of nature as a portal for artistic expression.

Perhaps our most common means of experiencing nature is through recreation—hiking, climbing, rafting, skiing, kayaking, fishing, camping and a myriad of other wilderness activities that all entail being in outdoor settings. These activities often immerse us in our sport while we enjoy a deeper awareness of and connection to nature. The closer to the "element," whether it's the snow from skiing, the water from kayaking, the rock from climbing or the stream current from fishing, the deeper our immersion. We are transported

Andrés Edwards

Paddling a kayak immerses us in a state of being where we are aligned with the rhythms of nature.

into a different state of being in which we are attuned to the rhythms of nature.

Types of Nature Alignment

What is alignment with nature? It is being attuned to nature's subtle changes and activities. I had a friend who loved watching the "dancing trees." When I observe a tree's trunk and branches, even on a day with only a light breeze, I notice a slight dance. A similar dance occurs in the rustle of leaves, the flow of water, the flight of butterflies, the orbit of our planet and the motion of all the stars in the universe. When we are in alignment, we feel connected to other species and to the natural world and from this connection arises a caring and love for nature. When we are in alignment we have a clearer sense of how we fit into the fabric of life, giving us an opportunity to redefine our role, if we choose to, as land stewards rather than as consumers or resource exploiters.

What happens when we are not aligned? We feel separate from the fabric of life that sustains us. This separateness manifests in a sense of isolation from other species. Nature becomes merely a set of "resources" that keep us alive: the air we breathe, the water we drink, the soil in which we grow our food, the timber and fish we harvest and the minerals we mine, all of them simply objects for our consumption. Instead of seeing ourselves as members of a larger community of species, we become detached exploiters of nature's bounty, disconnected from what nature provides. We also lose track of the knowledge of what keeps us alive, including where our water originates, where our food is grown and where our electricity is generated. All of these essential sources help us connect to the land. In a sense they are the threads that reconnect us to nature, and when we lose these threads we are adrift.

Some of the most complex environmental issues we face—climate change, loss of biodiversity, pollution—are exacerbated because we cannot clearly see the connection between our actions and their effects on the environment. We are not aligned with the rhythms and cycles of nature. How, for instance, does food that travels 1,500 miles impact the climate? How do large multinational fishing fleets impact the livelihood of coastal fishing villages? And what effects do industrial agriculture and unsustainable timber harvesting practices have on the health of the land and its capacity to replenish itself? These issues demand that we step back and reenvision our relationship to the natural world, become aware of the impact of our actions and then seek enduring solutions.

Since the majority of the world's population lives in cities and only a small percentage of residents from the developed countries are directly involved in activities such as farming, fishing, logging and mining, many of us are removed from daily exposure to nature. People living in inner cities with limited opportunities may rarely have a chance to travel to a wilderness area and see a starry night, a rushing river or an elk or a bear. For them, nature is an abstraction that may appear boring or even dangerous. I recall a friend telling me that when his family first moved from the city out to the suburbs his young children were afraid and had a difficult time falling asleep at night because of the sound of crickets, which they had never encountered before.

People who are more fortunate and have grown up with access to the wilderness see nature as a destination for recreation, contemplation or an opportunity to relax and leave behind the stresses of modern life. In these situations we are able to immerse ourselves in an activity or simply enjoy the natural surroundings and revel in being at ease with what is all around us. It's an opportunity to be present in the moment and relish the beauty we may discover if we look closely. It's also a chance to develop a contemplative state of mind.

Identifying the qualities in nature that we appreciate gives us insight into our character. How does nature, for instance, exemplify resilience, endurance, adaptation, thresholds, cooperation and interdependence? Maybe we see examples of these qualities in the branches of an oak tree, the flight of a dragonfly, a rugged coastline or a meadow in springtime. How do these qualities show up in our own lives? How can we fortify our own resilience, understand our thresholds and discover our interdependence with our family, friends and colleagues? Nature is the springboard for exploring our personal lives and provides a mirror for how we can navigate our social connections.

Nature also provides a timeline to give us perspective. Geologic epochs give us a sense of how species and the geography of the Earth have evolved over time. Changes in nature may take seconds or minutes from a lighting strike, a flash flood or an earthquake or millions of years as in the erosion of the Grand Canyon or the evolution of our own species. This long-term perspective provides a context for gauging the impact of our actions. Sometimes human activities such as the construction of cities or development of landscapes may appear to be long-lasting, but in the geologic scale 100 or 500 or 1,000 years are a mere fraction of time that may be a blip on the geologic radar. Taking a hike in the Grand Canyon, a forest or a city highlights the impermanence of life and the constant evolution taking place year after year, much of which we cannot detect on a day-to-day basis.

Internal and External Alignment

Aligning with nature has an internal and an external component. These two aspects are interdependent. Although they complement each other, they are essentially part of a whole. As philosopher Thomas Berry pointed out, "The outer world is necessary for the inner world; they're not two worlds but a single world with two aspects: The outer and the inner. If we don't have certain outer experiences, we don't have certain inner experiences, or at least we don't have them in a profound way."[5]

In learning about our internal alignment we may ask ourselves what rejuvenates us—exercise, yoga, walks, friends, gardening, reading, music? What brings us joy—time with our spouse, kids, grandparents, friends, a walk in nature? What brings us peace—meditation, reading, painting, solitude? What gives us purpose and meaning—pursuing our passion, being of service? What opens our heart—witnessing a compassionate gesture, an act of kindness, the love between people? What inspires us—a beautiful musical

Andrés Edwards

Identifying a tree species or understanding the carbon cycle or the migration of monarch butterflies gives us insight into how nature works.

performance, an athletic feat, overcoming a daunting challenge? What humbles us—a starry night, a hurricane, a redwood forest? What makes us playful—making a snow angels and sand castles, flying a kite? And how does nature help us with our internal alignment?

To examine our external alignment we may wonder what is our connection to our family, friends, community and country—family ties, social networks, volunteering? What links do we have to water, soil, food and the air—health, recreation, business? What is our connection to our bioregion, our city or town—cultural, geographical, professional, personal? And what are our ties to the mystery and beauty of the natural world?

These questions bring up the various ways in which we create our alignment with nature, namely: personal experiences, cultural ties, intellectual and emotional connections. On the personal level, Chandra Taylor Smith, an educator from the Audubon Society, recounts the story of a recently immigrated Mexican mother living in Baltimore who shared her desire for her children to learn about birds "because these same birds we see in the park here in the city may have been in my mother's back yard back in Mexico before."[6] This direct link between the birds' migration and her own family's journey to the United States speaks to her affinity with a bird species that reminds her of her culture back home. Perhaps she also identifies with how the birds and her family were able to survive the migratory journey.

In another instance, a group of African American women who were recently trained as naturalists and were leading a group through a park on the south side of Chicago suddenly stopped and one of them said, "I feel like doing the Beyoncé dance, because it makes me feel so good to be out here!" Another said, "I feel free and relaxed in a way that I cannot be at home, in the city."[7] Then they proceeded to dance freely and share their joy in feeling vibrant and alive in nature.

In these two instances the personal alignment is sparked by a connection to a familiar sight such as a bird species and a moment of inspiration because of the natural beauty around us. I recall a similar connection as a child in Chile and climbing around in a pair of tall eucalyptus trees in our backyard. Now as an adult living in Northern

California, seeing eucalyptus trees and smelling the scent from their leaves bring back instantly those fond memories. Being in nature often sparks a memory or a feeling that takes us back to another time and place.

On a cultural level, one of the strongest connections to nature stems from food. Traditional dishes from various cultures help us bond with our cultural roots. Thinking of our favorite "comfort food" often takes us back to a childhood surrounded by family traditions and cultural holidays. Whether it's a hamburger from America, moules-frites (mussels and French fries) from Belgium, tandoori chicken from India, borscht from Russia or pastel de choclo (pie with corn, chicken, beef and vegetables) from Chile, each dish links us to our ancestors, our land and our culture.

The origin of fruits, vegetables and nuts also provides a bridge to particular regions: avocado, corn and squash from Peru, Mexico and Central America; blueberries and sunflowers from North America; potatoes and quinoa from South America; broccoli, cauliflower and walnuts from the Mediterranean; eggplant and peaches from Asia; and coffee and watermelon from Africa.[8] Food acts as a cultural and biographical thread that ties us to nature's diversity and our agricultural practices over the centuries.

The intellectual thread of nature provides us with a methodology and language for interpreting the natural world. Learning the scientific names and understanding the processes of nature give us a powerful tool for making discoveries and sharing them with others. Identifying a tree species or understanding the carbon cycle or the migration of monarch butterflies gives us insight into how nature works. Becoming literate about the environment—developing ecoliteracy—builds a bridge for sharing our understanding of natural systems. This common ecological language has its limitations, yet it provides a baseline from which to discuss nature.

As an external tool for understanding the natural world, ecoliteracy supports an objective approach for defining our connections to nature. This ecoliteracy language is multicultural, intergenerational, evolutionary and at its root intellectual. Learning this language challenges us to understand the scientific concepts and terminology that are the building blocks of nature. The physical, biological and chemical topics covered through ecoliteracy lend us a perspective that is supported across the world through scientific inquiry.

Ecoliteracy provides a basic understanding that aids in the debates surrounding ecological issues. Understanding the basic terminology and the "operating instructions" for the natural world gets us all on the same page when it comes to discussing important environmental issues. Unfortunately much confusion and misunderstanding arise because of eco-illiteracy—not having a background that provides a baseline for having sensible conversations about environmental issues.

However, the ecoliterate mind still benefits from an emotional component. The emotional alignment with nature manifests through the arts and sciences. Whether it's nature art such as the work of Andy Goldsworthy, the nature photography of Galen Rowell or the work of hundreds of thousands of naturalists, educators and scientists making discoveries every day, they all help us explore our curiosity through the natural world.

The emotional link to nature comes through biophilia (see Chapter 6). This is the door that for many of us is first opened in childhood when we naturally live more in the present moment and are filled with curiosity and affinity for life. When we are mesmerized by a twig or a ladybug or a heritage oak, we develop a visceral bond with natural systems that is born from a feeling, not an intellectual pursuit, and remains with us for the rest of our lives. It is the force behind our love of a particular place, a kinship with a particular species and a sense of belonging to something greater than ourselves, which is life itself.

Natural Principles

An enduring alignment with nature is based on a set of bedrock principles: interdependence, stewardship, regeneration and compassion. Instead of thinking of them as lofty ideals, we can find creative ways to live them in our daily lives.

As humans we are dependent on water, air, soil and plants as well as the biodiversity of species to stay alive. These elements form part of an integrated system that can't be broken down into its component parts. As John Muir so eloquently pointed out, "When we try to pick out anything by itself, we find it hitched to everything else in the Universe."[9] We experience our external interdependence with the natural world through our land stewardship practices that can either enhance or degrade the health of nonhuman species. Our internal interdependence happens more subtly through our emotional connection to nature, our empathy for other species and our appreciation for the beauty and mystery of life.

As environmental stewards we must explore our responsibility to the natural world. The role we choose to play is ideally aligned with our temperament—thus we may express our values as an activist, educator, entrepreneur, politician or community leader. While stewardship emphasizes our role as an integral part in the web of life, the most tangible approach begins at the local level by taking care of our own backyard and our community's parks and open space areas. Then we begin to discern the connections between our home and our region, our state, our country and the world. In a wired world with a 24/7 news cycle it's easy to get immersed in stories from distant lands. Yet perhaps the most important question to ask is: where can we have the greatest impact? As stewards, what actions can we take individually and collectively that will make a difference? The answer may be as simple as participating in a neighborhood cleanup or a community garden project or installing a drip irrigation system to conserve water.

Regeneration is a fundamental process of life, a process of perpetual renewal. Our bodies are constantly regenerating—at a cellular level our cells renew themselves, starting with our blood cells, bone cells, muscle and skin cells—to keep us vibrant. Similarly, nature regenerates itself through fires, rain and sunlight, which bring new life to animal and plant species and their habitats. Internally, we regenerate and reinvent ourselves as our knowledge and opinions evolve throughout our lives. Just as plants and animals adapt to their changing environments, we adapt to our constantly changing life circumstances and progress through our roles as children, adolescents, adults, workers, parents and grandparents. Our regenerative qualities flourish when we're open to new ideas and fearlessly embrace the unknown, trusting that our resilience will give us the strength and stamina to move forward.

Compassion (see Chapter 7), which many social scientists believe is inherent in our being and can be cultivated through our life experience, provides a powerful way to align with the natural world. Studies are revealing that the vagus nerve, a cranial nerve that influences our speech, head motion, digestion and heart rate, also impacts our empathy, sympathy and compassion.[10] Similarly oxytocin, called the "cuddle hormone," plays a role in our social bonding and makes us more trusting, generous and empathetic towards others.[11] Our bodies thus have the key "ingredients" for being caring and compassionate.

But compassion is not only a human trait. Researchers are discovering compassion in the animal world. Chimpanzees, for instance, will take care of their ailing and elderly kin by helping them and getting them water. Using their tusks, elephants will come to the aid of an injured member of their group, and bonobos, known to be mostly peaceful in their behavior, will lick the wound from an injury they have caused someone else.[12] These actions illustrate the similarities that we share with other animals. Moreover, these types of actions from other species bring out feelings of empathy and compassion in ourselves. Our own compassionate actions toward animals range from the extreme measures, such as efforts to save whales stranded in sea ice, to the rescue of dogs and cats in urban settings.

Awareness and Humility

A successful alignment with nature involves developing our awareness coupled with a dose of humility. The first step in awareness is to "wake up"—to open our eyes, listen with our ears and open our hearts to how we fit into the tapestry of life that surrounds us. Being aware involves recognizing our role and responsibility as the most powerful species on the planet. How do we wield that power with respect and reverence for all other species? How do we protect nature's cycles, processes and ecosystems that allow life to thrive on the planet? And how do we ensure that our actions serve to create a livable future for all? These challenges are more easily tackled at the individual level where we investigate our personal motivations. Identifying what gives our lives meaning and purpose leads us to make choices that we can evaluate for their environmental impacts.

Awareness also requires slowing down and being still instead of worrying that we will "miss out" if we don't always accept the next event to fill up our schedule. Attuning our rhythm to the rhythm of nature ensures that we have the space and time to rejuvenate and reflect, providing context and meaning for many of the challenges we face.

Iain McGilchrist reminds us of the importance of seeing the connections in nature for therein lies meaning. As he points out:

> Meaning comes from connection and our brains are designed to attend to the world in two ways. [The left hemisphere] sees static, distinct, lifeless pieces, fragments that are then just put together to make up a kind of mechanical world.... The right hemisphere is the one that sees everything is connected, that nothing is ever static, that nothing is ever discrete and separate from other things. Modern physics confirms that the world is like that, and poetry and music have told us that since time began.[13]

Discovering these dynamic connections heightens our awareness of being an integral part of the universe and living a meaningful life. As the connections become more apparent, we are often humbled by the complexity and beauty of the natural world.

Humility emerges in different ways. I am humbled by the vastness of a starry night, the beauty of a sunset and the force of a rushing river. I realize how tiny and insignificant our planet is in the context of the cosmos. I'm also humbled by witnessing the remarkable landforms and species that inhabit our planet—from the snowy peaks of the Sierra Nevada mountains to the coastal redwood forests to the tiniest ladybug resting on a leaf. Geologic forces have been around for millions of years and species have evolved accordingly. Now in my comparatively brief lifetime I have a chance to experience them.

Humility is often highlighted during natural disasters such as earthquakes, hurricanes, floods and droughts, where we have limited control of the events as they unfold. We are often humbled and awed by the destructive power of the natural world. Perhaps an equal humility is required as we develop technologies for which there is uncertainty about their potential consequences such as new chemicals, genetically modified crops and nuclear energy sources. Making decisions fueled by greed and commercial interests can often lead to undesirable results. Applying the precautionary principle ("better safe than sorry") to technological "advancements" calls for our humility to step forward and to proceed with caution or not proceed at all. Embracing a sense of humility recognizes that we are not the only players on the world's stage and that we can still be overwhelmed by natural forces.

Natural Ingredients

What are the "ingredients" we need for aligning with nature? To start off, we might want several "cups" of awareness, stillness, reverence and compassion. Then we can add several "tablespoons" of: humility, awe, appreciation, wonder, pleasure, playfulness, observation, curiosity, creativity, solitude and respect. Add to that a "pinch" of flexibility for being open and grateful for the unexpected. These ingredients make up the recipe for a reciprocal relationship with nature in which we not only derive nature's benefits but also give back by renewing habitats we have destroyed through our short-sighted exploitation of natural resources.

QUESTIONS AND ACTIVITIES

- How have your outer experiences in nature influenced your inner experiences?
- What does being "aligned with nature" mean to you? What actions help you to be in alignment?
- What are the principles that would help you to develop an enduring alignment with nature?
- Find out if/how ecoliteracy is being taught in your community and offer to help spread its message.
- Recall a time that you felt a sense of unity with nature. How did it feel? What impact did it have in your life?
- How does nature renew your mind, body and spirit?
- What makes you curious about the natural world? How do you manifest your curiosity?

2 Awe and Beauty

The most beautiful thing we can experience is the mysterious. It is the source of all true art and all science. He to whom this emotion is a stranger, who can no longer pause to wonder and stand rapt in awe, is as good as dead: his eyes are closed.

— ALBERT EINSTEIN

The greatest beauty is organic wholeness, the wholeness of life and things, the divine beauty of the universe.

— ROBINSON JEFFERS

No synonym for God is so perfect as Beauty.

— JOHN MUIR

Credit: © joda/Adobe Stock

I RECALL VISITING PINNACLES NATIONAL Park (formerly Pinnacles National Monument) in central California on a summer's day in the early 1990s and experiencing a moment of awe that seems as vivid as when I was first there. A friend and I hiked for several hours through the hot, dusty trails and late in the afternoon we reached a high, rocky lookout point. The view from the rock outcrop was magnificent—on this clear afternoon we could see the valleys and neighboring mountains for hundreds of miles in every direction. As we scanned our surroundings we noticed a couple of turkey vultures taking advantage of the spiraling thermal winds and rising ever higher into the heavens. These two birds were soon joined by a third, fourth, fifth and sixth vulture, and they all flew with their wings spread wide, soaring effortlessly above us in circles. The juxtaposition of the beautiful rust-colored sandstone with the azure blue sky and the black wings of these birds left me in awe. Time appeared to be suspended. I had a feeling of reverence, insignificance, admiration and unity with all life. I felt a solidarity with humanity and nature. This was a moment of alignment with the pulse and rhythm of nature.

Awe is defined as "an overwhelming feeling of reverence, admiration, fear, etc., produced by that which is grand, sublime, extremely powerful, or the like."[1] We may be awed by nature, art, a human feat, a musical performance or the remarkable talent and skill of an individual. My experience at Pinnacles was infused with reverence and admiration. It seemed as if time was suspended and the turkey vultures made me feel small in the much larger scheme of what was occurring. There was also a feeling of unity, of being connected to something powerful.

In recent years, awe has been the focus of extensive research by psychologists. Beauty, by contrast, has been explored from the time of the ancient Greeks and in the last centuries by transcendentalists including Ralph Waldo Emerson and Henry David Thoreau. Awe and beauty act as catalysts that align us with the web of life.

Psychologists Dacher Keltner and Jonathan Haidt have discovered that awe has two distinct qualities: perceived vastness (the feeling of something greater than ourselves) and accommodation, the need to integrate the sense of something vast into our being.[2] Perceived vastness is often accompanied by a feeling of timelessness. In nature this can happen, for instance, when gazing at a starlit night, the powerful current of a river, the delicate patterns of maple leaves or perhaps the transformation of a caterpillar into a butterfly. These events suspend time because we are focused on the details of what is happening before our eyes. Accommodation involves creating a frame of mind in which we can integrate the event into our life experience.

Very often, our brains long to make sense of an event and want to quickly change our experience from the mystery of awe into the cerebral accommodation of awe so that we can categorize it. Perhaps the feeling of timelessness comes precisely in between mystery and accommodation as we look for a way to place the event in context. In this in-between phase some of the astounding characteristics of awe emerge.

Awe's Qualities

Awe has the capacity to connect us to others and make us behave in positive ways—what psychologists refer to as "prosocial" behavior. In a study conducted at U.C. Berkeley, a group of participants was asked to experience a majestic grove of eucalyptus trees—the tallest in North America—and feel its awe-inspiring impact; another group was asked to look at a nearby building. A few moments later, when someone "accidentally" dropped some pens on the ground, the group that had been gazing at the eucalyptus trees picked up more pens than the other group. The first group was more open than the other group to helping someone. They also felt "less entitled and self-important."[3] This desire to be of service to someone in need shows how the powerful force of awe can promote acts of kindness.

Awe-inspiring situations often lead to creative outcomes. When we focus outward and are curious and filled with wonder, creativity

Dave Alan

Gazing at the delicate patterns of maple leaves
suspends time as we focus on what is happening before our eyes.

ensues. Jason Silva's *Shots of Awe* video vignettes speak to the creative flow from the mystery and beauty of life. As Silva says in his video:

> You know Henry Miller says, "Even a blade of grass when given proper attention becomes an infinitely magnificent world in itself." You know, Darwin said, "Attention, if sudden and close, graduates into surprise; and this into astonishment; and this into stupefied amazement."
>
> That's what rapture is. That's what illumination is. That's what that sort of infinite comprehending awe that human beings love so much is. And so how do we do that? How do we mess with our perceptual apparatus in order to have the kind of emotional and esthetic experience from life that we render most meaningful? 'Cause we all know those moments are there. Those are the moments that will make the final cut.[4]

Jason Silva's impromptu, nonscripted pieces with accompanying music bring alive his passion for what makes life worth living. His expansive topics are beautifully packaged into short videos that evoke a yearning to connect with life. His work also shows how technology can be a gateway for encapsulating awe through the weaving of words, images and music. It's fascinating how he chooses places in nature such as the rugged coast of Northern California to expand his mind and come up with the narrative riffs that reveal the nuances of the topics he describes. The power of natural settings kindles his passion for that which lights up his and his audience's hearts.

Awe also has a positive influence on our health. Our body's cytokines, the chemical messengers that are part of our immune system, can help fight disease and heal us. Studies suggest, however, that people with "hyperactive cytokine response" are more susceptible to illness. The levels of cytokines appear to go down when we experience awe. Of all the positive emotions, awe causes the most significant drop in cytokines, leading to a reduced chance of illness.[5] Awe also make us happy and reduces our stress levels.

What propels us to seek awe-inspiring experiences? Perhaps at the root is the drive of our innate curiosity. Curiosity may leave us vulnerable to what is unfathomable and when we reach the cusp of the unknowable, we are awestruck. At this moment, we are beyond a cognitive state and simply experience what is happening with an open heart without language, labels and explanations. Our breath is "taken away," our eyebrows rise, our eyes widen, our jaw drops—and "WOW" is what we feel throughout our body. Then our brain kicks in and we attempt to assimilate what just happened. The sunlight at dawn, the magenta-colored evening clouds, the Cirque du Soleil's acrobatic feat, the emotive cello performance, the unexpected act of kindness, the powerful breach of a whale, the majestic flight of a condor, the emergence of a daffodil and the silver reflection of moonlight shimmering on ocean waves all instill awe.

Why do we need awe? It helps us see the world with a "beginner's mind" and fresh eyes. In making us feel small, awe expands our understanding of greatness. Witnessing the heroic feats of leaf-cutter ants makes us appreciate the remarkable strength, dedication and teamwork of this tiny species. Seeing the green flash of a sunset over the ocean connects us to the power of the ball of fire that makes life possible on Earth. Experiencing awe augments the social bonds that are innate to humans and have been an essential quality for the survival of our species for millennia.

As technology encroaches on our lives and we become "glued to the screen" of the smart phone, the computer, the TV or other devices, we get absorbed in mundane daily responsibilities. How do we retain awe in our lives? Perhaps we need look no farther than our immediate families and our backyards, where we may witness the kindness of a child, the bountiful cherry blossoms in springtime, the patterns of a sparrow's nest or the pelting of a summer's rain. These awe-inspiring events make us reach out to our neighbors and friends and strengthen the social connections that build stronger communities.

Beauty's Attributes

Just as awe motivates us to reach out and help others, beauty encourages us to care for nature. One of the most influential modern environmentalists was Douglas Tompkins, a successful entrepreneur, designer, farmer and activist who protected millions of acres in

Chile's and Argentina's Patagonia. If there was a common theme that ran throughout his life, it was beauty. As a designer, he had a keen eye for beauty that spanned from the clothing and equipment designs he led at Esprit de Corps and The North Face to the craftsman designs of the lodges in Patagonia and the wildlands conservation projects and numerous national parks he created in South America.

In his eulogy of Tompkins in 2016, Peter Buckley described a commitment to bridging the relationship between sustainability and beauty. As Buckley noted, Tompkins believed that "if we are to understand the widespread destruction of our current culture, we must attend to the loss of beauty. It's a zero-sum game. Doug's commitment was to reconnect beauty with life itself and with our efforts to sustain the world. Attend to beauty at every level, in every place, all the time."[6] As Tompkins himself said, "If anything can save the world, I'd put my money on beauty."[7]

Attending to beauty in nature calls on us to go beyond science, resource extraction and economic interests and to recognize nature's aesthetic value. The loss of biodiversity and habitats includes the loss of the beauty that is a component of all ecosystems. Beauty is an integral aspect of the fabric of life that we must attune to and protect. Beauty is the bedrock of Tompkins' legacy and remains an important pillar for advancing the conservation movement.

Beauty is a key driver of inspiration, emotional connection and an enduring ecological worldview. Transcendentalist Ralph Waldo Emerson described the ever-present thread of beauty in nature in his essay "Beauty," stating: "Nature is a sea of forms radically alike and even unique. A leaf, a sun-beam, a landscape, the ocean, make an analogous impression on the mind. What is common to them all,—that perfectness and harmony, is beauty. The standard of beauty is the entire circuit of natural forms,—the totality of nature...."[8] When we view beauty as a transcendent quality that permeates nature, the question of whether beauty is in the eye of the beholder or not is sidelined. Emerson adds, "Every natural feature,—sea, sky, rainbow, flowers, musical tone,—has in it [something] which is not private, but universal, speaks of the central benefit which is the soul of Nature, and thereby is beautiful."[9] Considering beauty as part of nature's soul provides an opportunity to treat beauty as a thread that connects us to the natural world.

Emerson's contemporary Henry David Thoreau was also inspired by the beauty in nature. He saw beauty as the precursor to meaning and relevance. As Thoreau pointed out in a journal entry, "With our senses applied to the surrounding world we are reading our own physical & corresponding moral revolutions. Nature was so shallow all at once I did not know what had attracted me all my life. I was therefore encouraged when going through a field this evening, I was unexpectedly struck with the beauty of an apple tree—The perception of beauty is a moral test."[10] Thoreau's moral test is less about good versus evil and more about recognizing the fundamental beauty that we are all capable of seeing in nature and incorporating into our lives. Beauty is the essential fabric of the web of life.

On occasion we may experience beauty mixed with awe and fear. Hiking in the woods in my hometown one afternoon I suddenly heard a rustling sound followed by hoofs pounding the hillside in rapid succession. Then from my left side came a pair of startled deer streaming downhill. They crossed right in front of me on the trail. At that instant I was startled and flooded with the beauty and awe of the moment mixed with some fear at being taken by surprise as they barreled by. After the moment had passed I began to think about whether the deer were running from a dog or other danger and if they had left behind others from their group. These mental "accommodations" seeped in as my mind tried to make sense of what had just happened, but the raw beauty and awe of this incident were what I had first experienced and what I recall most vividly.

The emotional connection to beauty occurs when nature evokes a passionate reaction that stirs us to our core. This emotional state may bring us tears of joy or a deep-seated connection with life. It may also be tinged with fear of the unknown and of what is to follow. As we surrender to the beauty of a particular moment, we are often enthralled.

In one of his most memorable passages, conservationist Aldo Leopold wrote, "Examine each question in terms of what is ethically and esthetically right, as well as what is economically expedient. A thing is right when it tends to preserve the integrity, stability, and beauty of the biotic community. It is wrong when it tends otherwise."[11] Leopold emphasized that economics alone should not guide us. "Beauty of the biotic community" has a very powerful impact on our emotional connection to nature and in turn will spur us into action on nature's behalf. Beauty helps us first to understand, then to care and eventually to love a landscape and become active participants in its protection.

What Emerson, Thoreau and later Leopold understood very clearly is that beauty is the constant that helps us support the vital web of relationships in an ecosystem. The beauty in nature is a key indicator of its integrity and vitality. When a habitat is damaged, as through overgrazing, development, pollution or deforestation, its beauty is altered. Preserving this beauty supports nature's services (such as clean air, clean water, healthy soils, pollination), which are the bedrock of a healthy ecosystem. Acknowledging the beauty of nature leads us to feel a kinship that expands our relationship to nature. In this way, beauty is the antidote to the commodification of nature.

Beauty challenges us to go beyond the scientific and monetary value of nature and cherish its "priceless" qualities: the sense of connectedness, social cohesion, happiness and relaxation that we derive from experiencing the beauty in nature. This shift moves us beyond the utilitarian worldview of natural systems into one that values nature's wholeness. This wholeness disintegrates when we unsustainably exploit, for example, oil, lumber, water and wildlife for their economic rewards. Beauty helps us reimagine our role in protecting the viability of natural systems.

Reinstilling Awe and Beauty

How do we reinstill awe and beauty in our everyday experiences? We may begin by slowing down, by taking a moment to be still and "smell the roses." We can encourage children and adult friends and colleagues to take a breath and reexamine the miracles that occur every day in our own neighborhoods: the intricate pattern of a robin's nest, the daffodil sprouting in spring, the ladybug crawling on a native bunchgrass. And beyond the natural world, we can admire an act of kindness between a mother and son, a compassionate gesture between strangers or an awe-inspiring work of public art. These are the little moments that bring joy and gratitude into everyday experiences. They make us feel alive and enhance our health and well-being.

Rather than dismiss technology as a barrier to experiencing awe and beauty, we can recalibrate how we integrate technology into our lives. We can design our "media diet" around healthy choices that support our connection to nature and advocate for teaching children media literacy to help them choose content that supports this connection. Taking a break from the 24-hour information overflow gives us a respite from the incessant chatter of popular culture and allows us to rediscover our own rhythm and be more open to the awe and beauty that surround us.

Awe opens the door to finding ingenious solutions to the problems we face. Marveling at the awe-inspiring characteristics of nature often propels our curiosity to delve deeper into the "how" rather than the "what" of situations. Investigating photosynthesis, for example, whereby plants take sunlight and convert it into chemical energy and produce oxygen, brings forth the mystery of a process that happens every minute of every day on Earth. Peeling back the layers of this ecological "onion" and understanding this process in depth allows us to imagine solutions to problems we may never have thought of. It may spur us to develop more efficient photovoltaic cells or devise a new method for storing energy from the sun or wind or tides. Staying in a state of awe fosters the conditions for creativity and solutions to emerge.

How does beauty fit into problem solving? Perhaps the answer lies not in focusing on beauty but in seeing how beauty manifests itself in a solution. As inventor R. Buckminster Fuller said, "When I am working on a problem, I never think about beauty but when I have finished, if the solution is not beautiful, I know it is wrong."[12] This leads us to gauging beauty as one of the measures of a successful solution. Perhaps beauty is so bountiful in nature because it has had billions of years of evolution to solve problems and devise solutions that are not just practical and efficient but also beautiful. Biomimicry (see Chapter 4) uses nature as a model to design enduring solutions to problems. These solutions can be seen in the physiological adaptations of animal and plant species; the ways species obtain food and water and design shelters; and how they devise hierarchies to manage their social groups. All of these solutions have worked through trial and error over millennia and acquired beauty as part of their evolution.

Beautiful solutions are often simple. The water and nutrient cycles in nature are examples of ways natural systems maintain healthy habitats. Over time, these cycles have evolved to maintain life on Earth without complexity or waste. As a mentor, nature can teach us how to regenerate life on the planet. We can learn how to cultivate a strong stewardship ethic in which we mimic nature's solutions without destroying the very systems we depend upon for our survival—clean water, clean air, healthy soil and abundant food.

When we experience awe and beauty our senses awaken and we feel connected to ourselves and to nature. I had an opportunity to experience the power of awe and beauty on a recent kayak expedition to Patagonia. One of our objectives was to document the changes that have taken place in the region over the last 40 years and advocate for its protection as a national park. Several friends and I paddled for two weeks into one of the most pristine and beautiful fjords I have ever seen. Located west of Puerto Natales, Chile, Canal de las Montañas is thirty miles long with steep sides embedded with hanging and tidewater glaciers that rise five to seven thousand feet. This area's wildlife includes sea lions, imperial cormorants, condors and multitudes of fish and shellfish.

As we reached the U-shaped end of the Canal late one afternoon, I experienced an awe-filled moment. The sun gradually disappeared behind the jagged peaks and dozens of shorebirds were feeding on the mud flats. A stillness and serenity in the air inspired a reverence for this remote part of the Earth. I felt like an honored guest invited by the residents of this pristine habitat to see the activities happening at the end of a summer's day. The mountains, glaciers, birds, sunlight, clouds and water all created a sense of vastness and timelessness, the

Andrés Edwards

Located in the south of Chile, Canal de las Montañas awakens our senses with its steep sides embedded with hanging and tidewater glaciers that rise thousands of feet.

essential ingredients of awe. The deep azure blue of the glaciers, the "rabbit ears" snowy peaks, the dense beech forests and the feeding shorebirds completed the scene with their compelling beauty. Awe and beauty intertwined to create a humbling experience.

Experiencing awe in nature challenges us to see with fresh eyes the natural processes that are constantly unfolding. Awe helps us tap into the miraculous events that move us and give meaning to our lives. By feeling small and even fearful in the face of an awe-filled moment, we learn who we are and how we fit into life on Earth. Beauty in nature challenges us to see the aesthetic bedrock of all life forms. Applying this revelation to all areas of our lives, we can then see the beauty in a butterfly, a musical performance, a work of art and even the trajectory of our life's journey.

QUESTIONS AND ACTIVITIES

- When was the last time you experienced awe? How did it feel?
- How do you integrate beauty into your daily life?
- Does technology detract from or enhance your experience of awe and beauty in nature? How?
- Think of beauty and nature and spend 30 minutes describing it through writing, drawing, painting, etc.
- Create a journal describing your moments of awe and beauty.
- How can you preserve beauty in your own backyard?
- What is the role of fear and awe in your life?

3 Health and Well-Being

The wilderness is healing, a therapy for the soul.

— Nicholas Kristof

"Healing," Papa would tell me, "is not a science, but the intuitive art of wooing nature."

— W. H. Auden

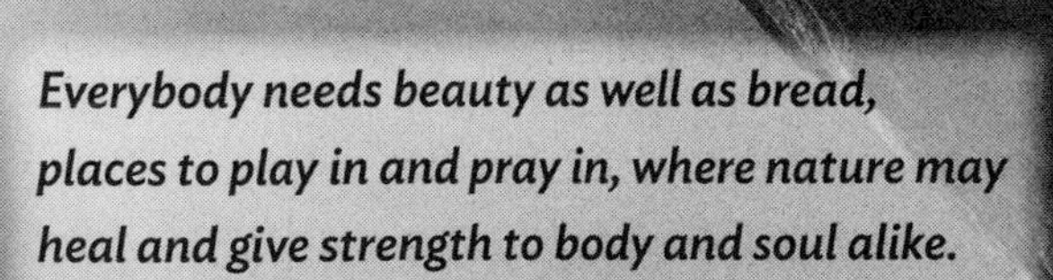

Everybody needs beauty as well as bread, places to play in and pray in, where nature may heal and give strength to body and soul alike.

— John Muir

Credit: © Konstantin Yuganov/Adobe Stock

During one of the most trying times in my life, when my marriage was coming to an end, I would venture daily into the forested area near our home and take the family dog for a long walk. Walking through the meandering fire roads gave me the space to reflect on my current predicament and what lay ahead. Having the dog as my companion was an added bonus.

During this difficult period, nature provided me with relief from the stress of a major life transition. The exercise, combined with the beautiful scenery and my dog's company, was my daily dose of soothing comfort. Gazing at the ancient redwoods reminded me of the many winter storms and changes they've weathered during their lifetime. Seeing a red-tailed hawk soaring above helped me to think of the need for a "bird's-eye view" for a broader perspective on my own circumstances. Witnessing the vegetation, ants, butterflies and squirrels mirrored to me that life is constantly evolving and adapting over time. Occasionally I would call a friend. Hearing their voice and support in the serenity of a natural setting also provided me with the strength to move forward through this challenging time.

Looking back several years later, I appreciate the healing power of my walks in nature. The wilderness gave me a place to reflect, discern, plan and exhale from the stress of the personal changes taking place. Taking the time to stop and look closely at the insects, the flowers, rocks and leaves rejuvenated my spirit and gave me renewed appreciation for how life is constantly unfolding around us. Even during the subsequent years as I adjusted to my new status, being in nature gave me a constant grounding that I cherish to this day. A friend who was also recently divorced mentioned that he made it a habit to buy himself flowers in order to bring beauty into his new surroundings. I welcomed the weekly ritual, which I still maintain, of having fresh flowers as a way to celebrate nature.

Healing Impacts of Nature

Nature serves as a refuge to inspire, reflect and heal. Studies reveal that being in nature has a powerful positive effect on the mind, body and spirit. The statistics on the health benefits for kids of being in nature are remarkable and in many ways not surprising. Outdoor activities increase physical fitness, raise levels of Vitamin D and improve distance vision; being in nature reduces ADHD symptoms; schools with outdoor education programs help students score higher in standardized tests and improve their critical thinking skills. Nature also reduces stress levels and enhances social interactions among children.[1]

These benefits also translate to adults. In adults, studies show that being in nature will speed the health recovery process, reduce blood pressure and lower the risk of cancer as well as lift people's spirits. In a classic study performed in a suburban Pennsylvania hospital between 1972 and 1981, patients who had a window view of deciduous trees healed from surgery much faster than those who viewed a brick wall. The patients with a nature view also received fewer negative evaluations from their nurses and took fewer pain injections.[2] High blood pressure, which affects one in three Americans, costs the US over $48 billion a year. A recent study, however, shows that adults can reduce their blood pressure by simply spending 30 minutes or more a week walking in a park.[3] In a study looking at the link between nature and cancer, people who took two long walks in nature over two consecutive days had an increase in their cancer-fighting cells, known as NK cells, of 50 percent and an increase in the activity of these cells by 56 percent. In addition, the activity levels of the cells remained high for a month.[4] These studies highlight the numerous ways that simply getting outdoors will benefit us psychologically and physically.

Some of the most interesting research on the connection between health and nature is coming from Japan. Walking and spending time in forests, known as *shinrin-yoku*, or forest bathing, is a popular form of preventive health care in Japan. Studies are now proving the health benefits of spending time in forests. Yoshifumi Miyazaki from Chiba University, Japan, discovered that going for a 40-minute walk in a cedar forest lowers the level of cortisol, a stress hormone, as well as blood pressure and supports the immune system more than a similar 40-minute walk indoors in a lab. Qing Li from Nippon Medical

Andrés Edwards

Being in nature can speed health recovery, reduce blood pressure, lower the risk of cancer and lift people's spirits.

School in Tokyo has shown that trees and plants emit compounds known as phytoncides that when inhaled give us therapeutic benefits akin to aromatherapy. Phytoncides also change the blood composition, which impacts our protection against cancer, boosts our immune system and lowers our blood pressure.[5]

Experiencing nature not only reduces stress but also improves our cognitive ability. Gregory Bratman from Stanford University and his colleagues enlisted sixty participants who were randomly divided into two groups: the first group took a 50-minute "nature" walk surrounded by trees and vegetation and the second group took an "urban" walk along a high-traffic roadway. The nature walkers showed cognitive benefits including an increase in working memory performance, "decreased anxiety, rumination, and negative affect, and preservation of positive affect."[6]

In a subsequent study, Bratman investigated the neurological mechanisms affected by being in nature by measuring the part of the brain (subgenual prefrontal cortex) that is activated by brooding. Our tendency to brood, referred to by cognitive scientists as "morbid rumination," often makes us focus on the negative aspects of our lives and can lead to anxiety and depression. Bratman and his colleagues found that the participants who walked in the quieter, wooded portion of the campus had lower activity in the brooding portion of their brains than those who walked near the busy roadway.[7]

The psychological benefits of being in nature are also affected by the biodiversity of the natural environment. As cities design urban green spaces, incorporating diverse vegetation and wildlife improves urban dwellers' health and well-being. A study in Sheffield, UK, surveyed the effects of different habitat types such as amenity planting, mown grassland, unmown grassland, scrub and woodland and monitored the butterfly and bird species in these areas. Participants showed an increase in psychological well-being in habitats with greater species diversity. As researcher Richard Fuller and his colleagues point out, "The degree of psychological benefit was positively related to species richness of plants and to a lesser extent of birds, both taxa where perceived richness corresponded with sampled richness." Additionally, "Our results indicate that simply providing greenspace overlooks the fact that greenspaces can vary dramatically in their contribution to human health and biodiversity provision. Consideration of the quality of that space can ensure that it serves the multiple purposes of enhancing biodiversity, providing ecosystem services (Arnold & Gibbons 1996), creating opportunities for contact with nature (Miller 2005) and enhancing psychological well-being."[8] Fuller's study suggests that the biodiversity in a habitat affects our well-being—the more species diversity, the greater the positive impact on our health.

As the scientific evidence mounts that immersing ourselves in nature increases our health and well-being, the question that arises is, why? The two most common theories that explain this phenomenon are the psychoevolutionary theory (PET) and the attention restoration theory (ART). Psychoevolutionary theory focuses on the human ability to have "positive built-in reactions to natural environments."[9] In essence, our positive connection to nature including low stress and high spirits has evolved innately as part of our species development over millennia. This theory accounts for nature's capacity for improving our well-being but doesn't delve into the cognitive impact of nature on our brain. For this aspect, we turn to attention restoration theory.

Attention restoration theory looks at the two main types of attention that humans employ: directed and undirected attention. Directed attention requires us to focus on a specific task and block any distractions that may interfere with it. For instance, when we are working on a math problem, or engrossed in reading a literary passage or in assembling or repairing an intricate mechanical object, our brains are totally dedicated to the task at hand, requiring our direct undivided attention. After we complete the task we often feel mentally fatigued or drained. Conversely, when we are outdoors, we may enjoy observing patterns or a sunset, clouds, flowers, leaves or a beautiful meadow, which call on our undirected attention. Using our senses to touch, see or smell in natural settings doesn't require a task-specific, problem-solving approach. Instead we can enjoy our experience in nature and be rejuvenated by taking in the sights and sounds at a relaxed pace. Undirected attention is easy to summon and maintain and leads to reduced stress and anxiety.

What about experiencing nature's healing powers through technology? Is it as effective as being out in the real deal? Studies show that when workers are given the choice of a windowless working station or a plasma TV display of natural scenes, they prefer the plasma option. This option improved their well-being and cognitive abilities. However, another study found that participants who had a view of nature through a window had a greater sense of well-being than those who simply had a blank wall; having a plasma TV "window" was no more restorative than a wall. So, as might be expected, views of nature are the most beneficial for our mental health, followed by photos or videos of natural scenes.[10] Medical professionals are catching on to the benefits of nature and incorporating in medical facilities architectural designs that include views of nature, images of natural scenery, natural lighting and healing gardens.

Nature As a Vital Supplement

Richard Louv, author of *Last Child in the Woods* and *The Nature Principle*, started a national discussion about the importance of nature in children's and adults' lives. He coined the term "nature deficit disorder" to highlight the negative impacts on children of spending less time outdoors and more time indoors, typically absorbed in their TV, computer, tablet or phone. Louv also speaks about the importance of the mind/body/nature connection, which he calls vitamin N (for nature). As he points out:

> Today the long held belief that nature has a direct positive impact on human health is making a transition from theory to evidence and from evidence to action. Certain findings have become so convincing that some mainstream health providers and organizations have begun to promote nature therapy for an array of illnesses and for disease prevention. And many of us, without having a name for it, are using the nature tonic.

> We are, in essence, self-medicating with an inexpensive and unusually convenient drug substitute. Let's call it vitamin N—for Nature.[11]

Using the nature tonic, or vitamin N, as an antidote for many of the ailments associated with modern industrial life reveals the importance of integrating nature into our daily lives. Since most of the world's population now lives in urban centers, parks and green spaces are becoming increasingly vital for our health and well-being. Even a 30-minute walk in a tree-lined area has proven physical and psychological benefits.

Louv helped to spark a national movement to get children outside in nature. In order to reverse the trend of children spending up to seven hours daily in front of screens, scores of organizations including the National Wildlife Federation and the David Suzuki Foundation are implementing innovative programs and resources for parents and schools. The National Wildlife Federation has set a goal to get ten million kids outdoors by giving parents resources to spend time with them outside and by working with schools and youth organizations to develop programs promoting unstructured time in nature. The David Suzuki 30x30 Nature Challenge encourages children and adults to spend 30 minutes a day outdoors for 30 days to kickstart a new trend, stating, "It is essential that we reframe our traditional view of nature as a place for leisure and sport towards one that emphasizes a full range of physical, mental, and social health benefits."[12]

In Scandinavian countries, the value of spending time outdoors is encapsulated in the word *friluftsliv,* which translates to "open air life." In Norway, Sweden and Finland *friluftsliv* supports a connection with nature that is incorporated as part of their cultural heritage. It means, for instance, kids playing outdoors and exploring the insects under rocks and logs or a bird's nest. In Finland teachers have competitive salaries, independence in their curriculum design, shorter school hours and plenty of time for their students to play outdoors. The success of their system, which blends work and outdoor play, has students repeatedly ranking near the top in academic achievement scores on a global scale. Playing outside is not merely an opportunity to rest and decompress but instead an important part of the learning process. As author Erik Shonstrom points out, "The central tenet of *friluftsliv* is the importance of entering into nature in an uncomplicated way. No Matterhorn ascent required—we're simply talking about kids playing in the woods, parks, and fields."[13]

For the general population in Scandinavia *friluftsliv* speaks to a deep appreciation of and interaction with nature. Originally introduced by Norwegian poet and playwright Henrik Ibsen, *friluftsliv* appeared in his poem, *På Vidderne*:

> In the lonely *seter* cottage
> My abundant catch I gather;
> There is hearth, a stool, a table,
> *friluftsliv* for my thoughts.[14]

Norway's majestic mountains, fjords, glaciers and vast pine and spruce forests beckon its residents to immerse themselves in nature. For some Scandinavians *friluftsliv* may entail spending time in a cabin nestled in the woods, for others a leisure afternoon in a secluded meadow, while others may take a hike to a mountain summit and enjoy the panoramic views of the wilderness.

Green Care

Whereas *friluftsliv* involves a more contemplative approach for incorporating the benefits of immersing ourselves in nature, green care emphasizes an active intervention by trained specialists aimed at providing physical, emotional and cognitive health and well-being. Using a variety of natural settings as well as domesticated animals is a powerful means through which these activities are carried out. Green care includes: social and horticultural therapy (working in garden environments), animal assisted therapy (using pets such as dogs, cats, rabbits and horses for therapeutic treatments), care farming (participating in farm activities as a therapeutic practice), green exercise (doing outdoor physical activities) and ecotherapy (nature-based therapy).

Social and horticultural therapy are used to help people with mental health issues develop social and motor skills. The American Horticultural Therapy Association supports organizations that implement horticultural therapy in garden environments, which includes designing therapeutic gardens. Horticultural therapy aims to "improve memory, cognitive abilities, task initiation, language skills, and socialization. In physical rehabilitation, horticultural therapy can help strengthen muscles and improve coordination, balance, and endurance. In vocational horticultural therapy settings, people learn to work independently, problem solve, and follow directions."[15]

Animal assisted therapy is a branch of animal assisted interventions, which more broadly encompasses using animals for rehabilitation purposes as well as companion animals. Each of these approaches may involve more or less therapeutic interventions with specific goals. People who have companion animals, for instance, have shown improved cardiovascular health, reduced psychosomatic disorders and fewer visits to the doctor, especially among the elderly. People who come into contact with animals such as horses, sheep, goats, rabbits, chickens, cats and dogs derive the benefits of feeling cared for, unconditional love and being less lonely. Closely related to animal assisted therapy, care farming aims to help participants suffering from depression, learning disabilities and drug addiction as well as disaffected youth and the elderly engage in farming activities such as planting and harvesting crops, horticulture, animal husbandry and land management practices.[16]

Although the physical benefits of exercising are well documented and accepted, the effects of exercising outdoors versus indoors are now showing some important psychological differences. The psychological benefits of green exercise include reduced stress and anxiety and improvements in self-esteem and mood as well as reduction in blood pressure and an increase in vitamin D from sunlight. Taking a walk outdoors in nature has also been shown to be effective in cases of depression.[17] Studies show that when we take a walk outdoors we exert ourselves more than when we do an indoor walk and, ironically, we feel like we're pushing ourselves less when we're walking outdoors. In a recent study, participants were asked to exercise for five minutes on a stationary bicycle while looking at: (1) a color video depicting a natural setting with green vegetation, (2) the same video with a red filter and (3) the same video without color. Even though the videos were otherwise identical, participants felt they had exerted themselves less with the color video and they reported being in a better mood than when they looked at the other videos. Perhaps from an evolutionary perspective we are better adapted for doing outdoor activities that have been part of our human development for millennia.

Having access to parks and nature has also been linked with longer life spans and a lower risk of mental illness in Japan, Scandinavia and the Netherlands.[18] I know that given a choice between walking, biking or jogging outside or in a gym, my preference (along with many others', I suspect) is to go outdoors. Some of the barriers to green exercise include: accessibility and maintenance of green spaces (some cities have greater access and better maintained parks and recreation areas than others), public safety and traffic congestion in getting to these areas. Parents are particularly concerned about safety, and kids are playing and exercising less outdoors as a result.

The psychological benefits of green exercise have given birth to an emphasis on the therapeutic aspects of nature through the expansion of psychotherapy known as nature guided therapy or ecotherapy. Linda Buzzell and Craig Chalquist point out that "*ecotherapy* represents a new form of psychotherapy that acknowledges the vital role of nature and addresses the human-nature relationship. It takes into account the latest scientific understandings of our universe and the deepest indigenous wisdom. This perspective addresses the critical fact that people are intimately connected with, embedded in, and inseparable from the rest of nature."[19] Ecotherapy incorporates many of the approaches already described. It also utilizes nature to facilitate personal introspection and reflection and finding our place within our community and our ecosystem.

Finding Our Place in Nature

What is the relationship between place in nature and our well-being? Some places may have a stronger calling than others for us. Many years ago during the introduction of a training that a group of us participated in with anthropologist Angeles Arrien, rather than say our names and our home towns, Angeles asked us, "What calls you the strongest: water, forest, desert or mountain environments and why?" For me it's mountains and water.

When I think about mountains and water I envision the numerous places that I've explored near my home to rejuvenate. The open space wilderness area just north of San Francisco includes the Mount Tamalpais State Park and the district watershed. It's rare and a blessing to have such a vast undisturbed wilderness area bordering a major urban center. Several years ago while hiking through the park I followed a deer trail that took me through some thick underbrush and ended in a panoramic vista of one of the reservoirs with the summit of Mt. Tamalpais in the background. Returning to this spot over the years has deepened my connection to this place and provided a gateway for reflecting and exhaling from the busyness of daily life. When I asked a friend who lives in Florida if she has a similar place in nature to go back to, she described a spot near her home where she loves to float effortlessly on the warm ocean surface. Finding our place in nature is at the root of discovering what keeps us healthy and vibrant.

QUESTIONS AND ACTIVITIES

- Identify that type of natural habitat that has a strong calling for you (desert, mountain, water...) and spend time there on a daily or weekly basis.

- Choose a green care activity (gardening, green exercise, ecotherapy...) and incorporate it into your weekly schedule.

- How do you integrate *friluftsliv* (open air life) into your life?

- Take a walk in a green space in your neighborhood (park, reservoir, wooded area...) and start a "walks in nature" journal to record how you feel.

- How has nature been a part of healing in your life?

- What role have animals played in your health and well-being?

- What activities do you do in nature with your family and with children?

4 Mentor and Provider

Credit: © quickshooting/Adobe Stock

After graduating from college I was seeking clarity for my next step. An opportunity came up to participate in a vision quest in the Utah desert, so I took it. About a dozen of us spent two weeks in a beautiful desert canyon in springtime. We knew that we would each be spending five nights and days alone fasting in a location of our choice. This was our time to reflect in a natural setting. I chose a rocky outcrop with magnificent views of a river as it meandered through the canyon.

Being away from modern day distractions provided me with the time and space to reflect and to observe in the moment what was happening around me from the tiny ants wandering in and out of a nearby mound, to the wind-sheared sandstone rocks that had been weathered through time, to the afternoon and evening thunderstorms that rolled through the desert. There was nonstop activity that became magnified only once I slowed down and took the time to observe it. I will always cherish the hardships, fears and joys that arose during this solitary experience. Nature was my ally and my mentor, reflecting back to me my thoughts and emotions so that I could examine them more deeply. Back with our group, I could revel in the similarities and differences of our experiences. Breaking the fast and nourishing my body again emphasized the essential role of nature as a provider of delicious food.

With 3.8 billion years of experience, nature has tried out many things and the ones that have succeeded are the ones that are still around and have stood the test of time. So whether it's species of plants or animals, microbes or habitats they, and we humans, have all adapted through millennia to be here today. A mentor is defined as "someone who teaches or gives help and advice to a less experienced and often younger person."[1] In this case, we humans are the "less experienced" and "younger person" with much to gain if we choose to be open, observe life around us and learn from nature, our mentor.

Another essential role of nature is its capacity as a provider. Nature provides us and all species with air, water, healthy soils, food, shelter, minerals and raw materials as well as beauty, danger, comfort, delight, awe and inspiration. Some of these are critical for our survival while others add to our human experience. As a young species we are still learning about the myriad ways that natural systems operate and how they benefit all life on Earth. Traditional ecological knowledge and modern scientific tools help us better understand our interdependence with the fabric of life that keeps us alive.

Learning from Nature

Perhaps one of the most exciting modern fields of study focused on learning from nature is biomimicry, derived from Greek: *bios,* meaning life and *mimesis,* imitation. In the 1950s, biophysicist Otto Schmitt, who coined "biomimetics," and medical doctor Jack Steele, who coined "bionics," were at the forefront of looking at nature for design inspiration. Both Schmitt and Steele were polymaths, sharing their passion and knowledge of different subject areas and discovering the connections between them. By the late 1990s, biologist and author Janine Benyus popularized "biomimicry" through her book *Biomimicry: Innovation Inspired by Nature.* Benyus defines biomimicry as "a new science that studies nature's models and then imitates or takes inspiration from these designs and processes to solve human problems...."[2]

Biomimicry challenges designers to look at nature with a critical eye and pay particular attention to the textures, shapes and processes that are present in nature but are often overlooked. When we look at nature from differing perspectives, new possibilities open up for innovative designs. One of the key questions designers might ask is: "How would nature solve this problem?" How do mussels hold onto rocks as waves crash upon them? How do birds and sea turtles navigate thousands of miles to their seasonal feeding grounds? Or how does a gecko effortlessly grip onto vertical surfaces and amble onward and upward?

Historically, humans have turned to nature for design ideas, guidance and inspiration. Leonardo da Vinci, for instance, carefully observed and made sketches of birds and bats to develop his concept of human flight and design his "flying machines"; he was later

fotogaby

How does a gecko effortlessly grip onto vertical surfaces and amble onward and upward? Biomimicry challenges us to see qualities in nature that are often overlooked.

followed by the Wright brothers. Textile designer William Morris looked to nature for his pattern designs. Swiss inventor George de Mestral went on a dog walk in 1941 and noticed that his clothes and his dog were covered with seeds from the burdock plant. Then using a microscope he discovered the hook-and-loop system the seeds used to attach themselves and this led him to the invention of Velcro, from the French "velours," meaning velvet, and "crochet," meaning hook.[3]

More recently, scientists and engineers at Duke University and the US Naval Academy have designed wind turbine blades that mimic the bumps, known as tubercles, on the front edges of whale fins. This new approach greatly increases the efficiency of the blades, reducing drag by 32 percent and increasing lift by 8 percent, making wind turbines capable of operating in areas where there is minimal wind. The design also has applications for propellers, fans and many other devices that use blades. Researchers investigating shark skin have discovered that the skin is made up of small, tooth-like scales of a material called dentin. These scales not only reduce drag but also create tiny vortices that prevent barnacles and other creatures from sticking to the shark. Emulating shark skin, scientists have designed a synthetic silicone skin that can be applied to boat hulls and reduces biofouling by 67 percent.[4]

Returning to the gecko, its remarkable ability to scale walls and ceilings is due to the miniscule hair fibers (called spatulae) on its feet that stick to a surface providing grip when pressed down and release when the pressure is off. Scientists are developing an adhesive that mimics the gecko's feet and will support almost a pound in weight; its application ranges from carpet tiles to climbing gear to medical equipment. In addition to products, biomimetic designs have been used in other industries such as architecture, where, for example, an office building in Harare, Zimbabwe, replicates the chimneys and tunnels found in termite mounds for heating and cooling its interior, using 90 percent less energy than conventional designs.[5]

Another way that we can learn from nature is through its ecological tipping points. We each have our own tipping points, from physical ones that we push during physically demanding activities to emotional ones that are tested in our daily lives. My personal growth work has helped me identify my tipping points by testing their boundaries. Looking at relationships, behaviors, fears and reactions brings greater personal awareness. On a cultural level, tipping points usually come unexpectedly and have a lasting impact on the course of history. The Pearl Harbor attack and widening of World War II; Rosa Parks and the US civil rights movement; Sputnik and the launch of the space race; Earth Day and the birth of the modern US environmental movement; Watergate and the lack of trust in the US government; the tearing down of the Berlin Wall and the end of the Cold War; and 9/11 and the age of global terrorism were all tipping points that defined profound cultural shifts. A question that naturally comes up is to what extent have the personal tipping points of world leaders had an impact on the cultural tipping points that have changed history?

A tipping point is "the critical point in a situation, process, or system beyond which a significant and often unstoppable effect or change takes place."[6] Examples from nature include: lakes changing from clear to turbid because of phosphorus and nitrogen runoff from nearby agricultural fields; deforestation that reduces rainfall, depletes soil nutrients and increases fire danger; and, at a global level, greenhouse gases causing climate change, which affects ecological, political and economic systems. Some tipping points occur very rapidly as a result of natural hazards with long-lasting effects. These include asteroid impacts, which have caused global extinctions, and volcanic eruptions, earthquakes and tsunamis, which have decimated habitats.

Nevertheless, nature has also taught us about resilience when habitats and species recover after these devastating events. Human impacts on nature, such as air and water pollution, chemical spills, species loss, toxic waste and oil spills also test nature's resilience.

Ecological scientist C.S. Holling first defined ecological resilience as "the capacity of a system to absorb disturbance and reorganize while undergoing change so as to still retain essentially the same function, structure, identity, and feedbacks." Holling and subsequent ecologists have identified four essential characteristics of resilience: (1) latitude: the greatest change that a system can undergo while still being able to recover; (2) resistance: how easy or hard it is for a system to change, its flexibility; (3) precariousness: the proximity of a system to its threshold or tipping point; and (4) panarchy: how the hierarchy of a system functions at "multiple scales of space, time and social organization."[7] We often witness resilience in nature when we return to an area after an event has occurred. When I returned to the scene of a forest fire in Northern California I was astounded by how the plants and trees and wildlife had come back after several years. The scars from the fire, including burnt tree trunks and shrubs, remained but overall the habitat had renewed itself.

Seeing how nature bounces back from adversity raises the question of how the characteristics of resilience in nature apply to resilience in our own lives. Some people demonstrate tremendous resilience in facing challenging situations while others buckle under the stress and never recover. What are the qualities that engender resilience in us? In her article "How Resilience Works" Diane Coutu points out the three main characteristics of resilience in people (which also apply to organizations): "a staunch acceptance of reality; a deep belief, often buttressed by strongly held values, that life is meaningful; and an uncanny ability to improvise. You can bounce back from hardship with just one or two of these qualities, but you will only be truly resilient with all three."[8]

The acceptance of reality calls for an honest assessment of what's going on in any situation, being neither in denial, which is an easy place to wallow, nor overly optimistic, which may blind us. When I fall into denial it's a comforting place to stay while not facing the truth. It's like being in a warm pool in wintertime. At some point it will be time to get out and face the cold air.

The second quality of resilience, finding meaning in life, is eloquently described by Victor Frankl in his seminal book *Man's Search for Meaning*, which describes his daunting challenge of surviving in a concentration camp in World War II. Frankl created meaning out of his dire circumstances by looking into the future. He envisioned

Andrés Edwards

This A-frame tree house has improvised ropes and counterweights to lift the heavy shingled roof.

himself teaching others what he learned about the psychological aspects of surviving such an ordeal. He found meaning in a tragic experience and created a worthwhile reason to stay alive. As he pointed out, "We must never forget that we may also find meaning in life even when confronted with a hopeless situation, when facing a fate that cannot be changed. For what then matters is to bear witness to the uniquely human potential at its best, which is to transform a personal tragedy into a triumph, to turn one's predicament into a human achievement."[9]

Resilience's third quality, the ability to improvise, comes up often when something unexpected happens and we need to go to plan "B" or sometimes "C" or "D" until we find a solution that works. I remember building a tree house with my sons, following drawings that called for an A-frame design with both sides of the roof fanning out as openings. With the shingles on, the roofs became too heavy to lift open. My sons' improvised solution: use two ropes with counterweights to make the roofs easier to lift. The plan worked beautifully. We devised a simple, creative, elegant solution rooted in improvisation. When we accept the reality of any situation we're facing, find meaning in our lives and innovate creative solutions, we nurture our growth and development and become more resilient.

Nature also learns through trial and error. Over the last 3.8 billion years it has experimented and evolved to flourish with species that have survived over millions and in some cases billions of years. What doesn't work hasn't survived. Among the oldest species are the nautilus, jellyfish and sponges, which are over 500 million years old. In the plant kingdom, the ginkgo tree and the cinnamon fern date back 70 million years. Cyanobacteria started producing oxygen through photosynthesis for the Earth's early atmosphere almost 3 billion years ago.[10] These species, which have stood the test of time, have gone through a trial and error process not unlike what we humans go through when we test to see what works and what doesn't. Making mistakes in our trial and error process gets us that much closer to discovering what works. In gardening, for instance, we experiment with the amount of sunshine, nutrients in the soil, species diversity and watering schedules. All these variables affect the productivity and vitality of the plants we grow, and through trial and error we learn what is most effective.

Cycles and Milestones

In addition to learning from nature's tipping points and from trial and error, we can also learn how nature takes advantage of cycles, how plants and animals attune to the cycles of the seasons for their survival. The periodical cicadas (genus *Magicicadas*) of North America, for instance, comprise numerous "broods" with a lifecycle that lasts either 13 or 17 years. Why 13 and 17 years? Well, these prime number years are designed to minimize the chances of being completely decimated by birds, reptiles, fish, even humans. By emerging in massive numbers, to the tune of 1.5 million per acre, cicadas utilize a strategy known as predator satiation, whereby their predators can have an "all you can eat" feast that still doesn't obliterate the species.

In spring when the soil temperature reaches 64 degrees Fahrenheit cicadas crawl to the surface. Adults live for four to six weeks, during which time the males make a deafening mating call to attract females. After the females lay their eggs on live twigs, the nymphs emerge, burrow underground and survive on the juice from plant roots for the next cycle of either 13 or 17 years. Then they emerge once again to complete a cyclical event that has been going on for millennia.[11]

The cycles of the cicadas and other species of plants and animals reflect our own cycles and stages in life. Our human cycles range from our internal body clock, which regulates functions such as our sleep/wake cycle, to the overarching birth/death cycle. In the interim we live through stages and milestones that define our human experience: graduations, marriages, divorces, birthdays, awards, performances, honors and career shifts as well as health issues, accidents and changing locations.

One of my greatest opportunities for personal growth and change was working as part of the news team for a local television station in Seattle, Washington. I had recently moved to Seattle and didn't know the area particularly well and taking this job opened a window wide into people's lives. I never knew what lay ahead until each morning when we were assigned a story—a shooting, a robbery, a car accident or a grand opening, a city council meeting, an election or a sporting event. In a strange way covering the news was a fast-forward, condensed version of witnessing other people's milestones. There were some tragic moments and some beautiful and awe-inspiring moments—from the demise of passengers in a seaplane accident near Seattle to the eruption of Alaska's Mt. Augustine volcano. My own milestone, after witnessing these events firsthand, was learning to adapt to unexpected situations with grace and respect.

Adjusting to a new situation requires a learning curve, in nature as well as in our personal lives. Evolution has its own learning curve whereby species evolve and adapt over time. Plants and animals have refined characteristics for their mutual survival. The Joshua tree (*Yucca brevifolia*), for instance, has adapted to survive in the hot, arid Mojave desert of the southwestern US. Its survival depends on the yucca moth (genus *Tegeticula*), which is the only insect that has developed special mouthparts for pollinating the flower of the Joshua tree. The Joshua tree depends on the yucca moth for its reproduction and the moth's larvae depend on the seeds from the Joshua tree for their survival.[12]

Our learning curve ranges from artistic expression to agriculture, science, governance, transportation and technological advancements, which tell the story of the evolving human experience. But at the bedrock of these astonishing breakthroughs lies our interdependence with nature—from the air we breathe to the water we drink, the food we eat, the shelters we build and the energy we consume. As a species, our ingenuity has given us remarkable success in learning to adapt to a range of habitats and climatic zones. Our interdependence with the Earth's living systems is similar to the interdependence of the yucca moth and the Joshua tree.

Our social mutualism plays out in all the services and amenities that keep our communities running. Our trash collection, water, food, education, employment, transportation and healthcare are only some of the basic interdependent systems that we as individuals depend upon in our community for our own survival and well-being. Many of these services depend on resources that are gathered from thousands of miles away. On a personal level, our mutualism shows up when we rely on family and friends for their assistance in child rearing, caring for grandparents and providing moral and emotional support. In these mutually dependent activities, everyone reaps the benefits.

Nature As Provider

Ecosystem services are the numerous benefits we receive from nature. These benefits include clean air and water, healthy soil, pollination, disease regulation and flood and erosion control. Additional benefits include natural resources such as oil, wood and fish, as well as climate regulation, now destabilized by climate change. Among the cultural benefits we derive from ecosystem services are recreational and educational opportunities, aesthetic pleasure and spiritual fulfillment.[13]

How do all these benefits relate to our well-being? Since the linkages are not always clear, consistent or easy to measure, this has been a challenging question to answer. The United Nations Millennium Ecosystem Assessment investigated the connection between ecosystem services and the five main aspects of well-being: security (personal safety, security from disasters); basic material for a good life (food, shelter, adequate livelihood); health (feeling well, clean air and water); good social relations (social cohesion, helping others); and freedom of choice and action (achieving what an individual values doing and being).[14] In some cases there are strong connections between ecosystem services and human well-being, such as the links with food, water, energy and basic material for a good life; other connections are weaker and less well-defined, such as with recreational and spiritual benefits and good social relations.

An underlying moral question that affects the link between ecosystem services and our well-being is what is the "intrinsic value" we place on nature. To what extent will our decisions be guided by what's best for our well-being versus what's best for ecosystems? These choices don't need to be mutually exclusive, but conflicts often arise when, for example, an endangered species impedes a housing development or a costly environmental regulation aimed at protecting the environment affects the bottom line of a business enterprise.

Nature's ecosystem services provide the essential life-support system for all species including us. Similarly, when we aspire to a life of giving back and being of service, we flourish and increase our health and well-being. Volunteering, kindness and being grateful, compassionate and empathetic all impact our well-being in a positive way. Just as nature enriches and nourishes us through its ecosystem services, these actions benefit us in our health, growth and development.

Nature's limits allow it to thrive. When these limits are exceeded, an ecosystem gets out of balance, with effects that ripple throughout the biotic community. Many of our actions cause these limits to be exceeded. Recognizing these limitations supports not only the health of ecosystems but also of ourselves. As entrepreneur Paul Hawken points out, "It is precisely in the discipline imposed by the limitations of nature that we rediscover and reimagine who we are and what economy truly means."[15] Financial, time and resource limitations help to set the context for our creative pursuits.

My experience in working on exhibits for museums and visitors' centers often focuses on the most effective way to convey information. What's the best way, for example, given a set budget and available resources, to show visitors what's unique about their bioregion? What stories need to be told? Answering these questions within the limitations of the institution should focus not only on making do with less but also on the assets of given project. In the case of a small natural history museum, this means highlighting their in-house collection, their research programs, the benefits of their location and their relationship with the local community. All of these "limited" resources can support imaginative solutions and a successful outcome.

Nature's limits and its elements and cycles provide useful guidance for our lives. Trees, for instance, provide a home for many species, bear fruit and support their structure during winter storms with their deep roots and flexible branches. Similarly, we have our own roots, which are the values and principles that help us navigate through difficult times. Just as water often flows continuously through rivers and tides, seeking the path of least resistance, following a natural path often helps us identify our talents. Rivers also illustrate a sense of beauty and timelessness. As environmentalist David Brower said, "We must begin thinking like a river if we are to leave a legacy of beauty and life for future generations."[16]

The unseen often becomes visible within a larger context. The force of the wind becomes apparent only with a background of an object such as a flag or leaves flying. We understand our actions better when we examine them within the larger context of our character and motivations. Mountains convey a sense of strength and stability. We can emulate these enduring qualities in our relationships and projects that can withstand the test of time. And the beauty of flowers reminds us of the beauty that surrounds us in nature and the kindness and generosity of others. Flowers also depend on the generous acts of pollinators much as we depend on the love and kindness of others to flourish and thrive.

In addition to being a mentor and provider, nature also serves as a powerful source of inspiration. From the time of the drawings of horses and deer in the Lascaux caves of France 17,000 years ago, nature has inspired us to tell stories that describe the human experience. Poets (Hafez, Rumi, Rainer Maria Rilke, Robinson Jeffers, Robert Frost, Emily Dickinson); writers (Henry David Thoreau, Ralph Emerson, Rachel Carson); painters (Vincent van Gogh, Paul Cézanne, Claude Monet, Winslow Homer); architects (Frank Lloyd Wright, William Browning); photographers (Ansel Adams, Galen Rowell, William Henry Jackson); and environmental artists (Andy Goldsworthy, Nils-Udo) have all turned to nature to inspire their creative genius.

We learn from nature's teachings when we slow down and practice stillness. When we take the time to carefully notice the wind carrying the clouds, tree branches dancing or ants or bees going through their daily actions, we are rewarded with glimpses of how nature, our mentor, works. Observing these events happening all around us also reminds us that perhaps as "young" students we might benefit greatly if we take a humble approach in our quest for learning. After all, we humans are only one among 30 million other species, most of which we know nothing about.

QUESTIONS AND ACTIVITIES

- Find one of your favorite places in nature and stay there for 30 minutes recording your observations in a journal.
- How has nature been a teacher for you?
- How has nature been a source of inspiration for you and how have you expressed it?
- Do you think of yourself as a resilient person? If so, how has being resilient benefited you?
- What poets, writers, painters, photographers have inspired you?
- What have been the defining milestones or tipping points in your life and how have they shaped you?
- Design and implement an environmental art project in your backyard or a nearby park.

5 Nature's Intelligence

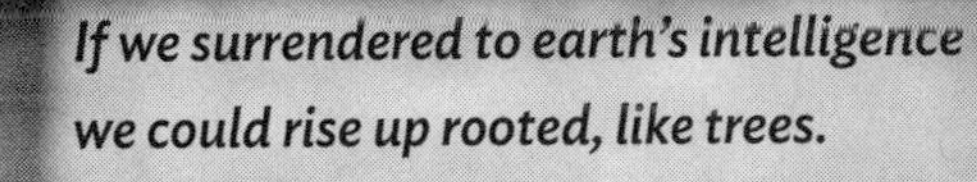

If we surrendered to earth's intelligence
we could rise up rooted, like trees.

— Rainer Maria Rilke

The major problems in the world are
the result of the difference between how
nature works and the way people think.

— Gregory Bateson

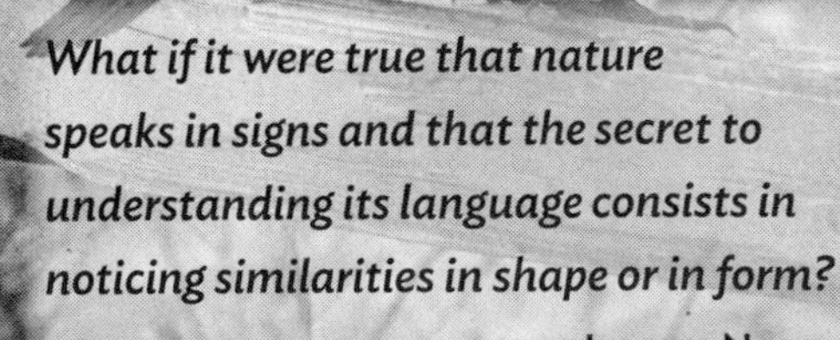

What if it were true that nature
speaks in signs and that the secret to
understanding its language consists in
noticing similarities in shape or in form?

— Jeremy Narby

Credit: © Patricia Thomas/Adobe Stock

My experience with nature's intelligence comes from being suddenly awakened by the curiosity of animals. Curiosity and exploration often go hand in hand with intelligence. In the first instance, I was sleeping outside on a wooden deck beneath a plum tree when at dawn in the middle of a dream I was awakened by a deer. I woke up when the buck's nose rubbed mine. Needless to say, I was startled when I looked up and stared right into his eyes and his antlers. He took a step back and we looked at each other before realizing what had happened; then he casually walked away. I've held on to my vision of the buck's eyes and antlers as my first sight that day.

In the second instance, I was with a group of a dozen teenagers attempting to summit Mt. Rainier in Washington State. We had climbed halfway up the mountain and were staying at a shelter overnight. Because of the limited space, I was sleeping on the floor. Once again, while asleep I felt a scratching sensation inside one of my ears. I thought it was a pesky mosquito. Then I rose and saw it was a mouse! It had been gnawing in my ear—for what I'll never know.

Both instances were startling and unexpected. And yet the deer and the mouse were doing what any curious animal would do: checking me out as an intruder in their space, as I would do if I saw something that piqued my curiosity. While our consciousness makes us distinct from all other species, to discover the intelligence in nature we need to be keen, humble observers who seek out the subtle signs that show how nature manages to adapt and thrive, from microscopic fungi to the largest marine mammals. In the broadest sense the planet also has intelligence. According to the Gaia theory, the Earth behaves as an intelligent living organism.

Intelligence of Gaia

When we look at nature's intelligence on a planetary scale, we see how the Earth moderates its gases, nutrient cycles, seasons and climate. Atmospheric gases including carbon dioxide, methane, hydrogen sulfide and oxygen are constantly being regulated by global geochemical systems. While some of these gases have fluctuated over millennia, others such as oxygen have stayed relatively stable for millions of years. These gases provided the conditions for life to emerge and to continue to thrive on Earth.

Within ecosystems, Earth's intelligence manifests in the way subtle changes in sunlight, moisture, nutrients and temperature affect the organisms in these communities. As David Abram points out, "Whether we speak of a whole range of mountains, or of a small valley within that range, in every case there is a unique intelligence circulating among the various constituents of the ecosystem—a style evident in the way events unfold in that place, how the slow spread of the mountain's shadow alters the insect swarms above a cool stream, or the way a forested slope rejuvenates itself after a fire."[1] These forms of intelligence define the biotic communities that live in diverse habitats ranging from dry desert to alpine, forest, shoreline, river and ocean environments. Many of these processes (rainfall, water cycle, carbon cycle, vegetation growth...) are constantly happening and we can measure them with our scientific instruments, or we can fine-tune our senses and our observation skills to notice these changes over time.

During one of my jobs as an environmental educator, I had the opportunity to spend an autumn in southern New Hampshire. Practically every day from late August through December I witnessed the seasonal changes happening in a 700-acre New England landscape that included forest, meadow, lake and bog habitats. Working with middle school children, we documented the changes occurring around the edges of the lake and the maple tree leaves turning from green to bright yellow, orange, red and eventually brown as the days grew shorter and colder. This was a unique opportunity to see the late summer, fall and early winter seasonal transition in one place. We noted how the deer, beavers, squirrels, foxes, hawks and butterflies prepared for winter and in many instances migrated south to warmer climates. Living in one place over time sharpened my observation skills and I gained a greater appreciation for the intelligent natural processes—some quite visible and others less so—happening all around us every day.

Why Nature's Intelligence Matters

As humans, we have transformed the planet more than any other species through deforestation, agriculture, urban centers, pollution, mining, fishing. We are now in what many scientists call the Anthropocene epoch, marked by the extensive human impact on Earth's ecosystems. Through our development, we have also changed the Earth's climate, putting us and all species in peril. As we discover the intelligence of nature we see that we are not the only "intelligent" beings on the planet (given the current state of the world, some would argue that humans are far from that). Perhaps one of our greatest challenges as members of the dominant species is to expand beyond our anthropocentric viewpoint and see the world from the perspective of other species. There are numerous examples of other plant and animal species that have their own intelligence, cleverness and capacity for thinking and making decisions.

The more familiar we become with other forms of intelligence, the more empathy and compassion we will generate toward other species. Knowing more about their remarkable abilities makes us appreciate them more and hopefully care for them. The "glamour" species such as dolphins, otters, elephants, polar bears, pandas, butterflies, redwoods, orchids and ospreys grab much of the attention and resources for their protection, while the rest remain virtually unknown. We need to understand the broader make-up of the web of life and the critical role that all species play in it. We can begin by slowing down and becoming familiar with the species that are our neighbors living in our own backyard.

Observing nature's intelligence is a huge opportunity for us to learn to live in harmony with the natural world. We humans have a lot to learn if we choose to do so: how to build efficient shelters, gather food, produce energy, store water, heal from disease and much more. A humble approach to understanding how nature operates will provide us with a proven manual.

Nature's Designs

Philosopher and psychologist David Stenhouse describes intelligence as "adaptively variable behavior within the lifetime of the individual."[2] This view points to the adaptability of ecosystems and their organisms. Particular patterns in nature describe smart designs that are economical, efficient and beautiful. The nautilus, for example, maintains a proportional symmetry in its shell as it grows into a spiral. This mathematical sequence, based on the Fibonacci numbers, also applies to the radial symmetry of snail shells, sunflowers, artichokes, pineapples, ferns, tree branches and numerous other living organisms. Nature's patterns have an inherent intelligence that allows organisms to flourish and it has proven itself through trial and error for millennia.

The hexagonal shape of honeycombs exemplifies the optimal design for bees to store the greatest amount of honey in the smallest surface area using the least amount of energy. In fact, the bees' engineering feat led to the hexagonal honeycomb conjecture, first proposed by Roman scholar Marcus Terentius Varro in 36 BC and finally proven in 1999 by US mathematician Thomas Hales. Hales' mathematical formula proved Varro's hunch that a hexagonal tiling grid pattern is the most compact and efficient way to divide a surface into equally shaped areas with the least total perimeter. A hexagon, rather than a square or equilateral triangle-shaped structure, is the most geometrically compact and space-efficient way to lay a grid on a flat surface without gaps.[3]

Beyond the honeycomb shape, let's look at the intelligence of the bees themselves. An experiment conducted by Martin Giurfa at the University of Toulouse, France, discovered how bees are able to distinguish between abstract symbols. At the entrance to a maze, Giurfa painted the color blue. Further down, the maze splits into two paths: one of them was painted blue and at its end was a sweet solution; the other was painted yellow and had no reward at its end. The bees consistently chose the blue pathway. When Giurfa used vertical and horizontal lines instead of colors, the bees were also able to choose the correct symbols painted at the beginning and at the

iStock: Riorita

The hexagonal shape of a honeycomb is the optimal design for storing the greatest amount of honey in the smallest surface area using the least amount of energy.

branching of the maze. With their miniscule brains, bees are capable of distinguishing symbols, a remarkable capability for learning. As Giurfa points out, bees "go into nature equipped with instinctive information, which is not rigid, and which they can forget or put aside on the basis of personal experience, meaning to say on the basis of learning."[4]

Butterflies have the ability to see a wide spectrum of color. With a brain no larger than two millimeters, they can see better than humans in some respects. Using ultraviolet photoreceptors, they can identify flowers with ultraviolet spots that indicate they contain pollen. They also see ultraviolet stripes on female butterflies, which plays a role in their courtship. Similar to bees, butterflies make decisions about which flowers to collect nectar from and what type of nectar (more or less viscous) to collect. Based on these activities, butterflies appear to be thinking and making choices. As entomologist Kentaro Arikawa, who has dedicated his career to studying butterfly neurology, points out, "I believe that there must be some primitive form of mind in these [butterflies], or the ability to think in things. I don't think that a simple chain of reflexes is sufficient to explain the whole thing."[5]

Animal Intelligence

Some of the most intriguing behavior occurs when a species acts as a group. Swarm intelligence describes species such as ants, fish, birds or caribou that demonstrate the power of the group rather than any single individual to make smart decisions. Scientists have identified three distinct characteristics of swarm intelligence: decentralized control; response to local cues; and some essential rules followed by each individual. In the case of ants, for example, when an ant finds food it leaves a chemical called a pheromone on its trail to alert others in the colony. As more ants discover the food using that same path they drop their pheromones on top of the original scents. In this way the optimal and shortest path toward a food source has a stronger and fresher scent than paths that are less desirable. Similarly, individual birds flying in a flock make sure they stay close to their neighbors. Each follows the direction of the flock. There is no single leader commanding the group; instead the birds are free to change their position within the group and collectively warn each other of a potential threat through their sudden motion. They also search for food using everyone's keen eyesight rather than relying on any single individual. This same intelligent approach is used by a school of fish and a herd of caribou.[6]

Wildlife biologist Karsten Heuer and his wife, filmmaker Leanne Allison, witnessed the swarm intelligence of a Porcupine caribou herd in Alaska as it was being preyed on by a wolf. In Heuer's words:

> [W]hen the herd was on the move it looked very much like a cloud shadow passing over the landscape, or a mass of dominoes toppling over at the same time and changing direction.... It was as though every animal knew what its neighbor was going to do, and the neighbor beside that and beside that.... As soon as the wolf got within a certain distance of the caribou, the herd's alertness just skyrocketed.... Now there was no movement. Every animal just stopped, completely vigilant and watching.... [Then] the nearest caribou turned and ran, and that response moved like a wave through the entire herd until they were all running. Reaction times shifted into another realm. Animals closest to the wolf at the back end of the herd looked like a blanket unraveling and tattering, which, from the wolf's perspective, must have been extremely confusing.[7]

The wolf unsuccessfully chased several caribou until it ran out of steam and the herd escaped. In this, as in other swarm maneuvers, every caribou appeared to know exactly when and where to run in order to confuse and collectively avoid falling prey to the wolf.

Researchers are applying their knowledge of swarm intelligence in the natural world to solve complex decentralized problems. These range from determining the most efficient truck routes for deliveries, to managing search engine results on the internet, to sorting entries on websites such as Wikipedia, to improving efficiencies in airline operations. Southwest Airlines, for example, is using an ant-based simulation model to improve its departure times by having airplanes remember the fastest gates and avoid the slowest ones at Sky Harbor International Airport in Phoenix. The benefits of collective intelligence are being recognized as a powerful approach to solving complex business and social problems. As Thomas Malone from MIT's Center for Collective Intelligence points out, "No single person knows everything that's needed to deal with problems that we face as a society, such as health care or climate change, but collectively we know far more than we've been able to tap so far."[8]

One of the most popular research areas of bird intelligence is focused on corvids, the family of birds including crows, ravens, jays, magpies and nutcrackers among others. Corvid studies have astounded scientists since in many ways their brains reflect some human characteristics. Crows, for example are able to recognize and remember human faces for years. In one experiment at the University of Washington, a researcher wearing a caveman mask trapped and banded crows. When the "caveman" who had abducted the crows returned, the crows remembered the incident and relentlessly pursued and dive-bombed him. Even after 10 years, when the original crows in the study had died, the new generation of crows attacked the caveman, showing that the experience and knowledge of this event was passed on from the older birds to their offspring. Crows and jays are also known to have "funerals," where they gather around their deceased kin, cawing together. Crows, along with chimpanzees, dolphins, elephants and fish, are capable of making tools to gather food. Corvids have also been known to remember dozens of hiding places where they store their food and can return to them many years later. In some instances, jays will recache their food to insure that it is securely hidden when they know they are being watched by crows.[9]

One of the areas that has expanded in recent years is research into the cognitive similarities between humans and birds. Research by neuroscientist Nicky Clayton has focused on crows' ability to think about the future and their "theory of mind," or the capacity to reflect on one's own mental state—such as desires or interests—and extend it to someone else's. In essence, this is an integral part of empathy. Clayton has also conducted experiments showing that jays have "episodic memory," or the ability to remember the past as well as plan for the future. In one study crows were able to find a specific food based on their memory of where it was stored. In the subsequent "planning for breakfast" experiment, scrub jays were allowed to store pine nuts inside trays in different compartments or "rooms." The study showed that the jays chose to store three times more pine nuts in the "breakfast" room than in the other rooms. They were thinking ahead, looking forward to enjoying the pine nuts the following morning.[10]

Dogs have long been trained to sniff the odor from people who have different types of cancer. A recent discovery found that pigeons are also capable of detecting cancer, and instead of using smell they can recognize the cancer patterns from slides. At the University of Iowa, researchers trained pigeons to look at slides and select a blue rectangle when they saw benign tissue and a yellow rectangle when there was a malignant tissue. In this experiment the pigeons that trained with specific slides were correct 87 percent of the time; those that had never seen the slides before were correct 85 percent of the time. Following a two-week training, pigeons also had an 85 percent accuracy rate for identifying mammogram slides that showed calcifications that can lead to cancer. Although pigeons aren't likely to be replacing pathologists any time soon, these experiments show their ability to discern subtle patterns and colors following a brief training period.[11]

Plant Awareness

Rather than focusing on whether plants are "intelligent," perhaps it's more appropriate to explore whether plants are aware. Are they cognizant of what's going on around them? Studies over the last decade show that they indeed are aware of their surroundings. As Daniel Chamovitz describes in *What a Plant Knows*, "They are aware of their visual environment; they differentiate among red, blue, far-red, and UV lights.... They are aware of aromas...know when they are being touched.... They are aware of gravity; they can change their shapes to ensure that shoots grow up and roots grow down. And plants are aware of their past: they remember past infections and the conditions they've weathered and then modify their current physiology based on these memories."[12] These remarkable attributes describe the different ways that plants know what is around them and change accordingly.

Similar to animals and humans, plants use their cells to sense and learn about the world that surrounds them. They do all of this without a brain. In seeking the best sunlight conditions, plants are capable of controlling their rate and direction of growth. They also seek out the optimal areas for growth. In one experiment ground ivy was shown to sense the nutrients in the soil and grow roots only in areas with good soil, while skipping sections with poor soil. Is the plant "thinking" about where to put down its roots or is this merely a physiological reaction? Perhaps, as plant biologist Anthony Trewavas from the University of Edinburgh says, the ground ivy is demonstrating its plasticity—its capacity to change its own structure and adapt its functions according to what it's experiencing internally as well as externally from the surrounding environment. The ground ivy is adapting to changing conditions, and rather than being instantaneous, these changes may take place over days or even months. As Trewavas points out, the changes "indicate that a lot of computation goes into the decisions which are actually made, otherwise plants would not dominate this planet in the way that they actually do."[13]

Plants use memory as a survival mechanism. The Venus flytrap, for example, is a carnivorous plant that senses when an unsuspecting insect is crawling over its leaves. The plant has a short-term memory that it uses to activate large hairs that detect an insect on its surface. When it feels its prey on at least two of its hairs, the trap snaps shut, encircling the insect. Because a large amount of energy is required to open and close the trap, the plant wants to make sure it catches a worthwhile meal. It gauges the size of an insect by the number of hairs the insect touches as it's crawling. If only one hair is touched the prey is not large enough and the plant's trap will not snap shut; however, when the second hair is touched the prey is likely big enough and the trap closes. Researchers have discovered that the first

Cathy Keifer

The Venus flytrap is a carnivorous plant whose short-term memory activates hairs that detect an insect on the surface of a leaf.

two hairs must be touched within 20 seconds for the trap to close. A small ant that touches only one hair is ignored.

The question of how the Venus flytrap remembers that its first hair has been touched by an insect and then transmits this information to its second hair has baffled scientists since the late 1800s. Then, several decades ago, Alexander Volkov and his team from Oakwood University in Alabama proved that electricity increases the plant's calcium levels that activate the trap. Volkov and his colleagues did an experiment in which they were able to close the plant's trap using electrical current without touching any of the hairs. So the Venus flytrap's short-term memory works through an electrical charge that spreads through its cells to activate the trap and dissipates after about 20 seconds. Its memory is indeed short-term, yet quite remarkable. Moreover, the electrical signals in the Venus flytrap are similar to the signals in other plants and animals including our own neurons.[14]

Whether or not plants hear the sounds of music has been explored and many scientific studies that have shown that music does not play any relevant role in a plant's development. However, scientists have been looking into the role of natural sounds, such as the buzzing of bees, to assess their impact on plant development. In the pollination process, bumblebees create a high-frequency vibration by activating their wing muscles without flapping their wings, releasing the pollen from a flower. In this way plants respond to vibrations without hearing them. As Chamovitz reminds us, "So just as deaf people can hear and respond to vibrations in music, flowers feel and respond to bumblebee vibrations, without necessarily *hearing* them. But conceivably, the sound of the vibrations could also affect the flower in some yet undetected way."[15]

Evolutionary ecologist Monica Gagliano has been investigating the cognitive abilities of the *Mimosa pudica* plant. In one experiment she noticed that when she dropped the plant one foot it would retract and curl its leaves. Then, after waiting for a month and dropping it gain, she found that the *Mimosa* didn't retract its leaves, thereby seeming to remember its past experience. In a subsequent experiment, Gagliano and her colleagues used a fan to "train" plants to grow toward the airflow to receive the "reward" of sunlight. Similar to Pavlov's dog experiment, in which the bell was rung and the dog salivated even when there was no food, Gagliano's plants grew towards the airflow when there was no sunlight. These experiments show how plants learn from experience and make choices. They also raise some complex ethical questions about the moral standing of plants given their brain-like abilities.[16]

Ultrasonic vibrations have been recorded in pine and oak trees during drought conditions. Roman Zweifel and Fabienne Zeugin from the University of Bern, Switzerland, have measured these vibrations, which "result from the changes in the water content of the water-transporting xylem vessels." Why release ultrasonic vibrations? They may be a survival strategy to warn nearby trees to prepare for a prolonged period of drought.[17]

Seeing, touching, smelling and remembering are all abilities that we recognize in our own lives. To appreciate these qualities in plants requires us to use a different and much more subtle lens that investigates these characteristics over time.

Slime and Mushrooms

Single-celled organisms such as amoebas, including true slime molds, also exhibit remarkable intelligence. Scientific studies by biologist Toshiyuki Nakagaki have shown that the slime mold *Physarum polycephalum* can solve a maze. When this blob-like slime mold is inside a maze it spreads out to fill all available spaces. When food is placed at the beginning and the end of the maze, the slime mold will withdraw from parts of the dead-end maze tubes and shrink to access the food in the shortest space available. Nakagaki describes how this brainless, single-celled organism demonstrates its "primitive intelligence" and challenges the notion that intelligence needs a brain. In this case, the single cell of a true slime mold repeatedly makes clever decisions that have been documented by scientists.[18]

Mushroom mycelium has also demonstrated some astounding capabilities. As mycologist Paul Stamets says, "When it comes to mushroom mycelium, the idea of natural, evolved intellect seems intuitive to me.... When I discovered some of the amazing responsiveness of mycelium, I was hard-pressed to think of fungal behavior as anything other than intelligent."[19] As a self-described mycelial messenger, Stamets is passionate about researching and educating the world about the genius of mushroom mycelium, including: fungi were the first organisms to evolve on land over 1.3 billion years ago; mycelial mat is the largest living organism on Earth, covering over 2,200 acres, and is only one cell wall thick; mycelium spreads across habitats and builds food webs that support other organisms; it sequesters carbon dioxide from the atmosphere and builds soil; it is phenomenal at "mycorestoration" and can break down diesel and oil spills and eliminate toxins in a matter of weeks; and it has saved millions of lives through the development of penicillin and more recently is being researched for fighting smallpox.[20] Unfortunately, scientists have identified only about ten percent of all the species of fungi. The rest are unknown, and they are disappearing before we even have a chance to catalog and learn about them.

Whether we are captivated by the characteristics of mushroom mycelium, true slime mold, bees, butterflies, ground ivy or the Venus flytrap, they all have developed ways to know their environment and make relevant choices that benefit them. While we may choose to call it intelligence or merely the capacity to know, these and all other species highlight the genius of nature at work at every scale in all ecosystems.

Although few of us have the scientific background to create experiments to study the intricate aspects of these species, as humans we are all born with curiosity and can formulate questions that intrigue us and whose answers shed light on how nature works. These questions help us understand the similarities we share with other species and cultivate our empathy toward all life forms.

Learning about the intelligence in nature also humbles us by showing that some of the most remarkable capabilities are exhibited by organisms that we may dismiss as insignificant. When we take the time to patiently observe the processes in nature, we discover the intricate design that holds together the web of life. Like the fingers in our hands, with their incredible flexibility and dexterity, each strand of the web has developed resilience and strength as it has adapted over time.

We are a very young species but we have been able to dominate all other species. We have learned to adapt to a wide range of habitats and to pass our knowledge down through generations. The lessons learned by our fellow species have enormous potential to help us find ways of living in balance with natural systems in order to have an enduring future.

QUESTIONS AND ACTIVITIES

- How does the Earth show its intelligence at a global scale?
- What are some examples that you've seen that show nature's intelligence?
- Choose a plant or animal species that interests you and discover how it shows its intelligence.
- Start a journal and record your observations of the activities of birds (including corvids) in your backyard or a nearby park.
- How has nature served as a mentor in your life?
- What is the relationship between curiosity and intelligence?
- Take a walk in nature and identify the various patterns in plants, animals, trees, hills, etc. How do they demonstrate nature's genius?

6 Kinship and Creativity

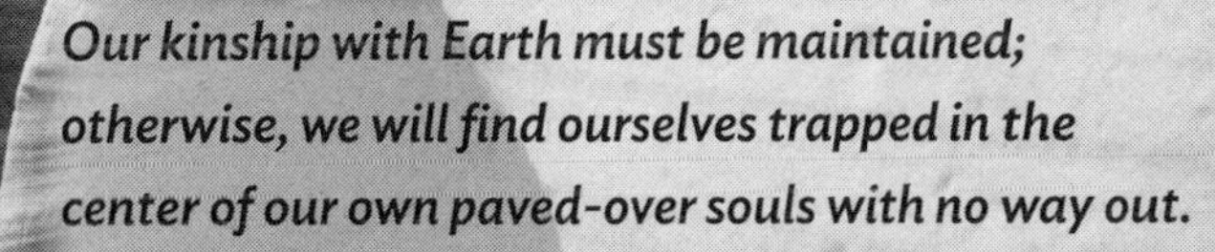

Our kinship with Earth must be maintained; otherwise, we will find ourselves trapped in the center of our own paved-over souls with no way out.

— Terry Tempest Williams

Wildness we might consider as the root of the authentic spontaneities of any being. It is that wellspring of creativity whence come the instinctive activities that enable all living beings to obtain their food, to find shelter, to bring forth their young: to sing and dance and fly through the air and swim through the depths of the sea. This is the same inner tendency that evokes the insight of the poet, the skill of the artist and the power of the shaman.

— Thomas Berry

It is the marriage of the soul with nature that makes the intellect fruitful, and gives birth to imagination.

— Henry David Thoreau

Our creativity often emerges in the most unexpected circumstances. While on a mountain biking trip with a friend in a remote region of Tibet, I noticed that the crank arm from my right pedal had suddenly come loose from the bottom bracket. As I attempted to pedal forward, the crank arm came off completely. We both knew this could potentially spell the end of our trip, since we had brought no spare nuts.

We decided to backtrack and after half an hour of searching for the nut in the dirt road, we came upon a Chinese farmer driving a tractor. I gestured to show him the specific nut that we were missing and miraculously he pulled out a nut and handed it to me. We tried it, but it didn't fit. Then he motioned for us to follow him, and after about fifteen minutes we came upon a roadside junkyard, the final resting place for worn-out road maintenance equipment. There were rusted-out spare parts from abandoned trucks, wheels, engines, earthmovers and heavy machinery everywhere.

My friend and I each got a wrench and started cannibalizing the equipment with the hope of finding a nut that would fit just right. Searching for the right nut in the junkyard became a creative game for me, and armed with a crescent wrench and a playful attitude I set out to find the nut we needed to solve our problem. After trying out dozens of nuts...Eureka! I found the perfect fit. Serendipity mixed with creativity saved the day, and the trip! I've kept riding that bike (now over 30 years old) and it makes me smile every time I think about the moment when I found the nut that is still holding the crank arm in place.

The hunt for the perfect nut to repair my bicycle is a mirror of our global quest to repair the ecological systems, many of which are either "broken" or in decline. Ocean acidification, desertification, deforestation, biodiversity loss and climate change are examples of ecosystem trends in dire need of creative "repairs." And just as I mistakenly thought I had a "quick fix" with the first nut the Chinese farmer showed me, there aren't "silver bullets" to solve the complex ecological problems we're facing.

Perhaps we need to creatively examine all the available options as we did in the junkyard, and then test them out to discover enduring solutions. And just as the perfect nut was hiding in plain sight, the solutions to our most pressing challenges are all around us and have been used by nature for millennia. We're simply being called upon to discover her "operating instructions" and to creatively adapt them so that we live in balance with natural processes. Luckily our love of nature enhances our creative juices, which are already producing many remarkable solutions.

What is the relationship between our love of nature and our creativity? Our attraction to nature often blends a mix of familiarity and curiosity. We may love to be near the ocean, a river, a mountain or a desert. At an ocean beach we may be attracted to the familiar sight of the seemingly boundless seascape; the sounds of the wind, crashing waves and shorebirds; the feel of the ocean breeze against our skin and the texture of the sand beneath our bare feet; the smell of the seaweed; and perhaps the "taste" of the salt and moisture in the air.

We may fall in love with the familiarity of a place that activates all our senses and with the curiosity that kindles our creativity. Quietly observing the patterns the wind creates on the ocean surface, the flight of pelicans skimming the waves, the brightly colored sea stars on the rocky shore and the tracks of crabs on the wet sand acts as a catalyst for our creativity. That "aha" moment when an idea is born often occurs when we're leisurely strolling in a natural setting and we "connect the dots," making an association we hadn't thought of before.

When we return over and over to a place in nature that we're familiar with, we have the added element of our history and memories (either positive or negative) associated with that location. Perhaps it was the place where we first met someone close to us, where we experienced an awesome sunset or moonrise, where we went on a walk during a time of grief or where we saw a coyote or a rattlesnake. All these experiences complete the mosaic of our connection with a place in nature.

Skalalitude

Our love of nature has been part of our evolution for millennia. The Salish peoples of the Pacific Northwest are particularly attuned to the connection of nature with where they live. They refer to this relationship as skalalitude. As James Swan describes it:

> Skalalitude means that when you are in proper harmonic relationship with the place where you live, the special places of power that are magnetic to you, and the many creatures of nature, then magic and beauty are everywhere. In a skalalitude state of mind, nature is a force for teaching and healing. To develop a skalalitude consciousness, Indian children are taught to learn to "listen with the third ear—the heart" by spending time alone in wild places, aided by supportive adult teachers.[1]

How do we allow our "third ear," instead of our analytical mind, to be at the vanguard of our connection with nature? Integrating nature into our lives is an innate aspect of our development that expands into a love for all living beings.

Social psychologist and psychoanalyst Eric Fromm first coined the term biophilia, which he defined as "the passionate love of life and of all that is alive; it is the wish to further growth, whether in a person, a plant, an idea, or a social group."[2] For Fromm the passionate love of life encompassed the conceptual, ideological and social aspects of life. Later, biologist E. O. Wilson refined and popularized biophilia, narrowing it to the environment's "living organisms," describing it as "the innately emotional affiliation of human beings to other living organisms. Innate means hereditary and hence part of ultimate human nature."[3] Our emotional affinity for nature is hardwired into our genes, perhaps dating back millennia to the days when we first roamed the African savannas. When did we first experience this innate connection to nature? For many of us, it goes back to our childhood experiences exploring the natural world.

In addition to childhood experiences, there are several other avenues that draw us into an emotional kinship with nature. James Swan describes five main pathways to our nature kinship: (1) intellectual knowledge, which arises from our yearning to learn about environmental issues and from the writings of naturalists such as Emerson, Thoreau, Carson and Leopold; (2) social justice, which encompasses injustice, inequality, poverty and suffering; (3) personal and public health, which involves taking action against health threats such as pollution, pesticides and toxins in the environment; (4) health and fitness, which draws on the connection between environmental and personal health including the promotion of organic foods, safe drinking water and staying physically fit through outdoor activities; and (5) emotional and spiritual experiences, in which our connection to nature brings us closer to the Divine.[4] To these five pathways, I would add creative expression as another essential avenue where nature acts as a catalyst for our creative pursuits. The creative expression inspired by the natural world may manifest as a piece of music or a dance, a painting, a sculpture or a photograph.

As Swan points out, "the simple accumulation of knowledge about ecology and pollution does not necessarily lead to a fondness for nature or an intuitive wisdom about how to live in harmony with nature."[5] Although we are bombarded daily with information about declining ecosystems, being informed doesn't mean we care or take action. Statistical facts and figures about environmental degradation can give us important knowledge but they often lead to our feeling overwhelmed and even hopeless. On the other hand, a photo or story (either positive or negative) that has an emotional component sometimes taps into our soul and galvanizes us to respond.

Such an instance occurred in 2017 when photographer Paul Nicklen from the environmental group SeaLegacy and a team from the National Geographic Society posted on social media a video and photos of a starving polar bear with the caption: "This is what a starving polar bear looks like. Weak muscles, atrophied by extended starvation could barely hold him up. Our @Sea_Legacy team watched as he painfully staggered towards the abandoned fishing camp from which we were observing and found some trash to eat—a piece of foam from the seat of a snowmobile, as we later found out."[6] This controversial, heart-wrenching post went viral and brought the

Bill Browning

Biophilic design brings the connection we feel with nature into the places where we live and work. The Art Aqua offices in Bietigheim, Germany, incorporate plants and water features as dividers.

face of climate change to a global audience. Although the incident ignited a short-term response, researchers have shown that enduring change calls for positive solutions that build on our affinity with nature.

Biophilic Design

Our connection to nature takes on added importance as the global population migrates to urban centers and millions of people live surrounded by the built environment. Social ecologist Stephen Kellert and environmental psychologist Judith Heerwagen defined biophilic design as "the expression of the inherent human need to affiliate with nature in the design of the built environment. The basic premise of biophilic design is that the positive experience of natural systems and processes in our buildings and constructed landscapes remains critical to human performance and well-being."[7] Biophilic design strives to bring the emotional and aesthetic connection we feel with nature into the places where we live and work. Rather than using sustainable design only to minimize environmental impacts, biophilic design focuses on bringing our connection to nature into the built environment, where increasingly people spend most of their time indoors. Natural light, plants, views and water are some of the features incorporated into biophilic design.

Architect Bill Browning and Terrapin Bright Green's study, *The Global Impact of Biophilic Design in the Workplace*, highlights the potential benefits of biophilic design in promoting well-being, productivity and creativity. Unfortunately, many workplaces lack the basic elements. Their survey of 7,600 employees from 16 countries showed that 58 percent of workers have no greenery in their offices, 47 percent have no natural light and 47 percent felt stressed in the workplace. On the upside: 67 percent felt happy in office spaces accented with green, yellow and blue colors. Workers with greenery and sunlight have a 15 percent higher level of well-being, a 6 percent higher level of productivity and a 15 percent higher level of creativity.[8]

Evidence of the biophilic effect of color in the workplace is still emerging. As Browning points out, the savanna hypothesis,[9] which describes our evolutionary adaptation to the landscape features of the African savanna, shows our preference for earth tones such as golds, tans and browns as well as blues (affinity to water) and greens (affinity to trees) found in these habitats. Blue and medium green are also associated with enhanced creativity. We're also drawn to some bright colors (affinity to flowers and fruits), so a few bright accent colors are preferred; red is associated with cognitive focus and vibrating color combinations may cause undesirable effects, even dizziness, so their minimal use is recommended. The color theory in interior biophilic spaces supports the notion that we are more likely to have a positive response to colors associated with resources that help us survive.[10] The important message is that biophilic design decisions including color choices, views, sunlight and greenery have an impact in how we feel and how productive and creative we are in the places where we live and work.

I've noticed the positive difference of working in offices that have wood, views, natural light and plants. I recall working for a design firm in a San Francisco building that had housed a coffee roasting business. As the firm got busier they expanded into an interior annex space with no windows. This was a challenging and unhealthy space to spend the whole day. And then one day the only exit door in the room jammed and a colleague and I were locked inside. My coworker began to feel claustrophobic as we tried to figure out the next steps. Luckily, eventually we were able to unjam the door. That experience brought home the importance of working in spaces with ample sunlight, views and greenery (and more than one exit!).

Studies of the benefits of biophilic design are reshaping how we think about the connection of nature and the built environment to our well-being. While designing green buildings focused initially on the efficient use of resources and the use of nontoxic, recyclable materials, we now focus on human-centered design: how architecture influences our behavior, health, productivity and creativity. This focus is crucial when designing hospitals that speed a patient's recovery by providing views of nature, schools that increase a student's learning performance with sunlit classrooms and offices that increase a worker's productivity and retention with plants, sunlight and views of nature. These design features have beneficial psychological and physiological impacts that make us feel more relaxed and help us thrive. The emphasis is on designing spaces for the people who live, learn and work in them. As Sandy Wiggins, past chair of the U.S. Green Building Council, reminds us, "Green building is not about buildings—it is about people."[11]

Creativity and Nature

Creativity in nature involves constant trial and error. This process identifies what works, which evolves to the next iteration, and what doesn't, which is left behind. Creativity requires a certain curiosity and open-mindedness. When we ask ourselves, "I wonder why this happens?" or "What's the best solution for this problem?" we enter a creative space where our imagination flourishes, often in a playful way. As psychologist Scott Barry Kaufman points out, "Creativity operates not as an efficient process but through a lot of trial and error; through a lot of mixing and matching seemingly contradictory personality traits and getting to really experience the full you—both your positive and negative emotions; going through set-backs and then trying to overcome them."[12] Creative solutions are often preceded by failed attempts that help us to home in on the best way to approach a problem.

Our unconscious mind plays an important role in the creative process. When we have a problem to solve and we decide to "sleep on it" our unconscious mind gets to work on coming up with options that often emerge as viable solutions. I remember working with a team trying to design an exhibit for toddlers that would be visually stimulating and involve active movement and water. One night I had a dream where I was lying down on my back at the bottom of a pond looking up at the surface. I could see fish swimming all around me and rays of sunlight coming through the water. We designed and built a space that included a kid-proof waterbed (one that was tough enough for use in surgeries on elephants), and above it we had a transparent water tank filled with live fish. Surrounding the waterbed were walls with buttons that toddlers could press and see water bubbles rising to the top. The space encapsulated the feeling of being underwater and the joy of jumping and twirling in the "water" while bouncing on a waterbed—an unconscious vision that is still being enjoyed by kids and adults!

What is it about nature that sparks our creativity? Perhaps nature helps us to slow down and relax enough that our senses are reinvigorated and new ideas surface. In his 1862 essay, "Walking," Henry David Thoreau described the positive impacts of immersing himself in a walk in the woods: "I think that I cannot preserve my health and spirits, unless I spend four hours a day at least—and it is commonly more than that—sauntering through the woods and over the hills and fields, absolutely free from all worldly engagements."[13]

Cognitive psychologist David Strayer's hypothesis is that being in nature allows the prefrontal cortex in our brain to calm down and in that rested state our creativity emerges. As he says, "If you've been using your brain to multitask—as most of us do most of the day—and then you set that aside and go on a walk, without all the gadgets, you've let the prefrontal cortex recover. And that's when we see these bursts in creativity, problem-solving, and feelings of well-being."[14] Strayer and colleagues Ruth Ann Atchley and Paul Atchley conducted a research study with 56 people who participated in four- to six-day Outward Bound hiking trips through the wilderness in Alaska, Colorado, Maine and Washington State. None of the hikers was allowed to bring a phone, computer or tablet. To evaluate the impact of nature on creativity, 24 of the 56 participants (a group of 30 men and 26 women) took a ten-question creativity test on the first day of their adventure and 32 took the test on the fourth day of their trip.

The test asked participants to identify a word that is related to a group of three words such as what word is related to cream/skate/water? (answer: ice). Participants who took the test the first day got an average of four out of the ten questions correct, while those who took it on the fourth day got an average of six right. The researchers concluded that "four days of immersion in nature, and the corresponding disconnection from multimedia and technology, increases performance on a creativity, problem-solving task by a full 50 percent."[15] Whether these results are due to the absence of technology or the immersion in nature remains a mystery but they may be a combination of both.

When the prefrontal cortex of our brain is active, it's focused on multitasking duties such as writing emails, answering phone calls and cognitive-intensive critical thinking and problem-solving tasks. These activities, referred to as "hard fascination," demand our full attention and often leave us feeling emotionally drained and mentally fatigued. When we're in nature, however, the prefrontal cortex has a chance to be restored. In this attentive yet relaxed state we allow our mind to wander freely and our imagination flourishes. Environmental psychologist Stephen Kaplan refers to this state as "soft fascination."[16] Our moments of insight emerge when we aren't thinking about anything specifically but simply taking in what's around us—water washing over a rock in a stream, the call of a bird, a butterfly spreading its wings, the flame of a campfire or cloud shapes in the sky. These natural sights and sounds captivate our attention without depleting us. We enter a space where unexpectedly an idea is born. Afterwards, it seems so obvious and we may ask ourselves, "Why didn't I think of this before?"

Soft fascination involves a beginner's mind that is curious, playful and flexible. Our creative juices begin to flow and we make associations we hadn't thought of before. Nature is full of opportunities to enter a creative trance as we observe the dynamic changes happening at every moment, changes that range from a tiny ladybug crawling over a leaf to a rolling thunderstorm crackling in the evening sky. And then, seemingly out of nowhere, the "aha" moment is born.

The good news is that we don't have to spend four days in the wilderness to spark our creativity. We can begin to restore our prefrontal cortex by taking a walk in an urban park, enjoying a summer breeze or even taking a break by looking at a tree outside our window—anything that lets us disengage from our attention-demanding tasks and have an opportunity to reflect.

ithinksky

Nature is full of opportunities to enter a creative trance as we observe the dynamic changes happening at every moment, ranging from a tiny ladybug crawling over a leaf to a rolling thunderstorm crackling in the evening sky.

Nature As Canvas

The natural world serves as inspiration for a multitude of artists who use nature as their canvas. Often called land art or environmental art, these works range from several miles long to tiny objects. Michael Heizner's 1970 land art project Double Negative, for example, involved digging two trenches in the Nevada desert, together measuring 50 feet deep, 30 feet wide and 1,500 feet long, which were opposite an existing natural trench, providing a contrast of two negative spaces, one natural and one artificial. Christo and Jeanne-Claude's famous large-scale art projects include: Floating Piers (2016), a floating walkway wrapped in yellow fabric on Italy's Lake Iseo; The Umbrellas (1991), where 3,100 umbrellas were displayed in an inland valley landscape in Japan and in California; and Surrounded Islands (1983), where 11 islands in Biscayne Bay near Miami were surrounded with a floating pink fabric.[17]

These large-scale land art projects help to stretch our imaginations by seeing patterns and colors juxtaposed in natural settings in ways that are unfamiliar. Yellow walkways on a lake, yellow umbrellas on hillsides and pink fabrics surrounding islands are "out of context," which is precisely the intent—to make us perceive and experience the landscape differently. They also serve as a metaphor for the creative process itself, where we're challenged to "connect the dots" in ways that we normally wouldn't think of doing.

On a smaller, more human scale, land artists like Andy Goldworthy, Patrick Daughtery, Walter Mason, Richard Shilling and Martin Hill and his wife Philippa Jones use twigs, leaves, stones, ice and other natural objects to create aesthetic patterns. Goldworthy's stone spirals, balancing rocks and ice sculptures; Daughtery's twig structures; Mason's leaf carvings; Shilling's stone sculptures and leaf patterns; and Hill's and Jones's circular sculptures rearrange natural objects to create beautiful patterns.[18]

Using sunlight, water, tides and wind these artists play with natural elements to accentuate the temporal aspect of their work, and indeed our lives. The warming of the day inevitably melts the ice sculptures, the ocean waves wash away the carvings on the beach

Andrea Pavanello

Giuliano Mauri's Cattedrale Vegetale (Tree Cathedral) relies on 42 beech trees that will eventually grow into a Gothic cathedral with a living vaulted ceiling.

sand, the wind blows away the leaf spirals and the incoming tide eventually engulfs the stone sculptures. Hill and Jones use photography as a means of capturing the ephemeral aspect of their work and to reiterate their vision for living in harmony with nature. As Hill points out, "My photographs are all that remain of the sculptures.... For me making this body of work is my way of connecting with nature to tell the story of the transition that is underway now towards a circular economy that emulates the way nature works."[19]

At the confluence of nature, architecture and art lie organic architects such as Marcel Kalberer from the Sanfte Strukturen studio in Germany and Italian artist Giuliano Mauri. These visionary artists and their colleagues use willow, reeds and trees to create living structures such as reed domes, willow palaces and tree cathedrals. Kalberer draws from ancient Sumerian reed techniques to build willow palaces that utilize bound reeds and branches for their support structure. Mauri's Cattedrale Vegetale (Tree Cathedral), completed in 2010 in the Lombardi region of Northern Italy, relies on 42 beech trees that will eventually grow into a full-fledged Gothic cathedral complete with a living vaulted ceiling. Made with fir poles and chestnut and hazel branches (as well as wood, nails and string), the cathedral has a nearly 7,000 square foot footprint is about 90 feet long, 80 feet wide and from 16 to almost 70 feet high.[20]

As artists and architects, Kalberer and Mauri blend their creativity with nature's vegetation to design living buildings that, like all life, continuously change over time. These examples of living architecture emphasize the use of recyclable natural materials that blend with the landscape. In addition, waiting for the plants to fully mature in order to complete the structures honors the cycle of life of which we are a part.

One of the most forward-thinking environmental art initiatives, known as the Land Art Generator Initiative or LAGI, was conceived in 2008 by architects Robert Ferry and Elizabeth Monolan. While living in Dubai and contemplating the rapid economic development in the United Arab Emirates they asked themselves, "What if the next 600-meter tall skyscrapers incorporated the latest in concentrated solar power and solar updraft technologies that could passively cool the interior while converting the sun's energy into power for an entire neighborhood?" and "What if our cities were populated by living buildings that functioned like canopy trees in a forest—converting the energy of the sun and the wind into electricity while passively regulating the environment?"[21]

Their quest for innovative ideas for designing beautiful buildings that produce clean, renewable energy without harming the environment led them to establish LAGI. LAGI's tagline, "Renewable energy can be beautiful," speaks to their intent to integrate beauty and energy production. How can we design clean energy systems that seamlessly integrate into the landscape, adding to its aesthetic value? LAGI's objective is to provide "a platform for artists, architects, landscape architects, and other creatives working with engineers and scientists to bring forward human-centered solutions for sustainable energy infrastructures that enhance the city as works of public art while cleanly powering thousands of homes."[22]

Ferry and Monolan created the first LAGI international design competition in 2010 and were astounded to receive hundreds of submittals from over 40 countries. Recent entries include LAGI 2014 Copenhagen, LAGI 2016 Santa Monica and LAGI 2018 Melbourne, all focused on producing "large-scale public artworks that could also generate utility-scale clean energy."[23] To encourage the next generation of artists and engineers, LAGI youth workshops inspire high school students to stretch their imaginations and critical thinking skills by designing energy solutions that support the STEAM (Science, Technology, Engineering, the Arts and Mathematics) education approach.

Beyond Nature As Resource

Over millennia humans have adapted to living in many diverse habitats and climate regions of the world. Through our ingenuity we have developed tools to exploit natural resources for food, energy and shelter but our affinity with nature goes much deeper. We have an innate, hard-wired, emotional bond with the natural world that supports our health and well-being, our happiness, our creativity and our artistic expression. The destruction of habitats and loss of biodiversity degrade the wellspring of inspiration we derive from nature.

Aligning our creativity with the health of natural systems provides a mutually beneficial outcome in which our technological advancements support rather than detract from the health of ecosystems and our love for nature becomes the driving force for its protection. Although the research into nature's impact on our lives is in its infancy, we know nature has profoundly positive effects. Our kinship with the natural world is the bedrock of the flourishing of our species.

QUESTIONS AND ACTIVITIES

- What is the role of nature in your creative pursuits?
- Do you have a place in nature that you're particularly fond of? What does this place offer you?
- What are the biophilic design elements in your home or workplace? If they are scarce, what can you do to make them more prominent?
- Take a nature walk and create a piece of environmental art using natural materials such as twigs, stones, leaves or other elements.
- Find a place in nature that has special meaning to you and write about your emotional connection to it.
- What are your first childhood memories of being in nature and how have they affected your life?
- Write a poem, make a painting or take a photograph of a place in nature that has a special meaning to you.

7 Compassion and Coexistence

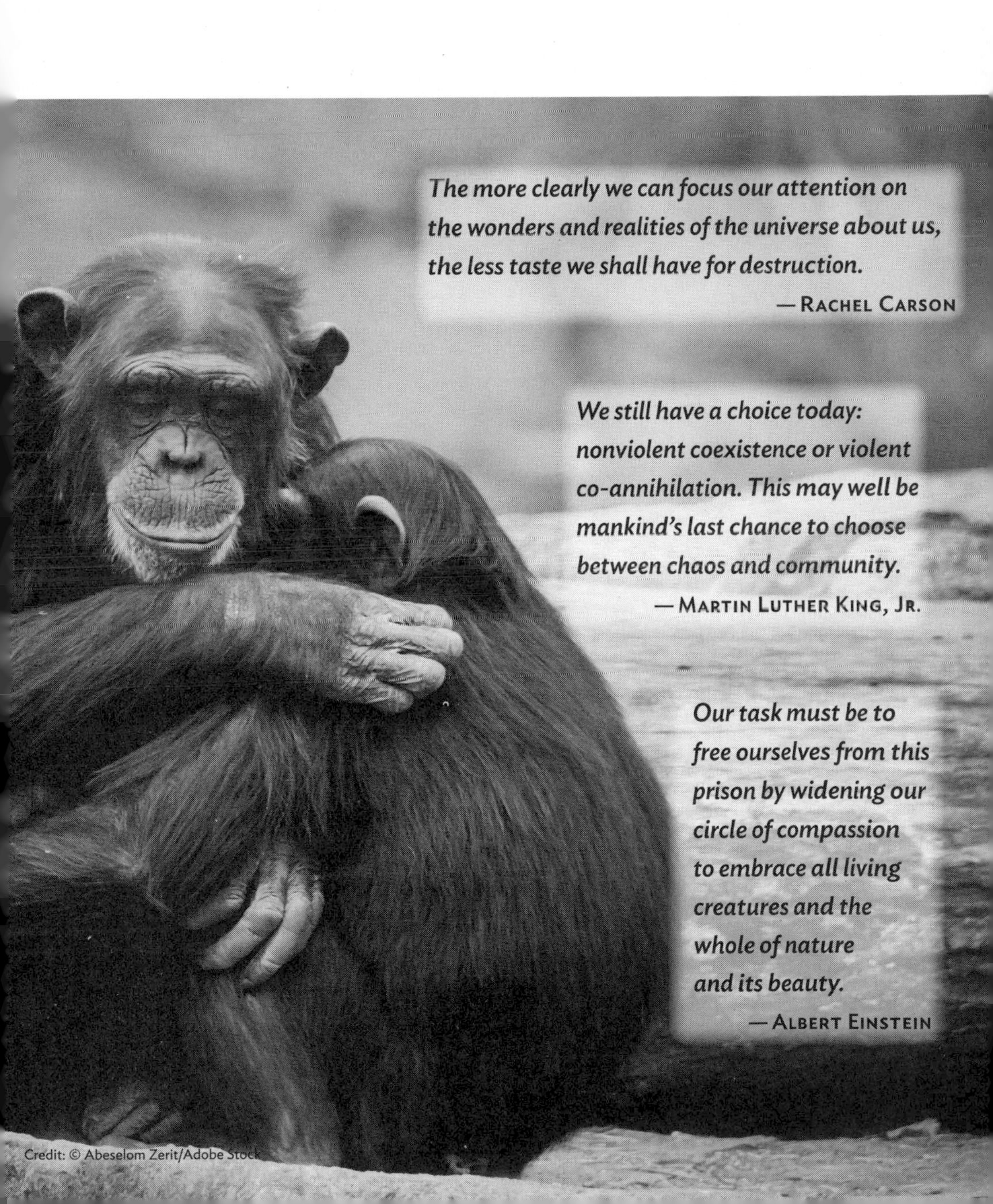

Credit: © Abeselom Zerit/Adobe Stock

One of my most vivid experiences of compassion for animals comes from my relationship with dogs. Although I have only vague memories, I was attacked by a dog in my early childhood and for many years was afraid around dogs. My fear changed when Monk, a mixed Rhodesian Ridgeback/Australian Shepherd rescue dog, came into my life. Monk had had a challenging start, growing up where he was abused. He barked at anyone wearing a hat, especially a baseball cap, so I presumed that his previous owner wore a baseball cap and I warned friends to not wear one around him.

Monk gradually recovered from his rough beginning and became my "healing dog." He taught me to have compassion for his early years and appreciate his canine qualities: his "singing" (long-winded howls), his protective nature, his impressive agility and his playfulness, kindness and patience, especially around children. Monk helped me reframe my perception of dogs and healed me from my own childhood trauma.

Cultivating Coexistence

During our childhood years our bonds with our pets are the primary relationship we have with other species. When we extend that relationship to wildlife, plants and habitats we're able to enlarge our perspective and appreciate that we are not alone. In fact, our survival depends on coexisting with a myriad of species. It benefits us to peacefully coexist with our "wild neighbors" since they help keep us alive. The "glue" that bonds us with nature is our cooperation, altruism and compassion, which have also been observed in other species. Perhaps reversing the environmental damage begins with reclaiming our connection to nature, which provides us with food, shelter, water, energy—in essence, our survival.

In the Quaker tradition, the principle of "bearing witness" instills a way for being in alignment with our core values. As Gray Cox points out:

> The guiding concern of people bearing witness is to live rightly, in ways that are exemplary. Insofar as they have an

> end they aim at, it is perhaps most helpful to think of it as the aim of cultivating their souls and converting others.... [G]enuine leadings all proceed from a common ground, spring from a unity which we seek and find.... This ground or source has a style, a style of conciliation rather than aggression, of inclusiveness rather than dominance, of organism rather than mechanism. Further, it has the character of care; it is respectful.[1]

The characteristics of "bearing witness"—to live rightly with care, respect, conciliation and inclusiveness—go beyond the Quaker tradition and have been the seeds of success for social change through, for instance, the civil rights, peace, women's and environmental movements.

Now, as we are being called to bear witness to the global ecological crisis, we ask ourselves how can we peacefully coexist with the rest of the species on the planet? Over the last century our actions have been degrading the soils, vegetation, oceans, biodiversity and even the climate of our planet. What is our responsibility as stewards to reduce our ecological destruction and show respect and compassion for the survival of other species? And how do we measure our progress on a global scale?

The United Nation's 2030 Agenda with its 17 Sustainable Development Goals in areas such as poverty, education, agriculture, energy, climate and terrestrial and marine ecosystems aims to "free humanity from poverty, secure a healthy planet for future generations, and build peaceful, inclusive societies as a foundation for ensuring lives of dignity for all."[2] Although there have been some positive developments over the last decade in areas such as reducing poverty and hunger, the overall rate of progress is not expected to meet the Agenda 2030 targets.

Biological Altruism

Perhaps the catalyst for changing from our current path of degradation toward a path of peaceful coexistence begins by deepening our relationships, which studies have shown fosters our innate altruistic behavior.[3] In humans, altruistic behavior is "when we act to promote someone else's welfare, even at a risk or cost to ourselves."[4] There are numerous inspiring stories of people's altruistic acts to help someone financially, or even risking one's life to save a child, a friend or a stranger. These altruistic acts are performed not only by humans but also by other animal species. In nature an organism behaves altruistically when "its behavior benefits other organisms, at a cost to itself. The cost and benefits are measured in terms of *reproductive fitness*, or expected number of offspring.... For the biologist, it is the consequences of an action for reproductive fitness that determine whether the action counts as altruistic, not the intentions, if any, with which the action is performed."[5] In the animal kingdom, altruistic behavior is rooted in the need to pass the genes on to the next generation. Whether it's bees or bats or birds or ants, they will sacrifice themselves for the survival of their kin.

Sterile worker bees, for instance, will sacrifice themselves for the queen's larvae. Honey bees will also sacrifice themselves when they use their stinger; after the stinger leaves their body it causes a fatal abdominal rupture. Some ant species hang upside down as "honeypots," providing food for their colony. One species of Brazilian ant, *Forelius pusillus*, protect the colony from night predators by sealing its entrance from the outside and dying from exposure to the cold overnight temperatures.[6]

Several animal species also practice "reciprocal altruism," in which the benefits of an altruistic action are shared. This is similar to an agreement such as "I'll scratch your back if you scratch mine." Female vampire bats, for example, which eat only blood and will starve if they don't eat for two consecutive nights, will share regurgitated blood with bats who haven't eaten. The benefits are mutual. A study showed that "when a fasting female bat had previously shared her food with other females, she received more total sustenance than a selfish one." Thus in this bat species altruistic behavior is rewarded by being taken care of when food is scarce.[7]

A classic study of reciprocal altruism is described by Berndt Heinrich, author of *Ravens in Winter*, who while hiking in Maine came upon a group of ravens making loud calls while feasting on a moose carcass. Calling attention to themselves seemed odd to Heinrich, since the birds would seemingly prefer to protect and not share their bounty. However, he noticed that the raucous group of ravens were juveniles feasting in a territory of mature ravens. The young ravens were calling other juveniles to join them in order avoid being displaced by the territory's dominant mature ravens. As Jeff Stevens, psychology professor at the Max Planck Institute for Human Development, points out, "True altruism...paying a cost to help another individual and never ever receiving any kind of benefit is not very common. It wouldn't make much sense biologically for that to happen."[8] In the case of the juvenile ravens and the vampire bats, behaving altruistically comes with a reward.

But what about humans? Are we intrinsically altruistic? Social scientists have shown that our instinct is to be cooperative. From when we're toddlers, we want to help others. The question posed by social researchers was: "Are people intuitively selfish but able to behave cooperatively with deliberate reflection, or are people intuitively cooperative, but capable of selfishness with further thought and reflection?" A Harvard University study based on the Public Goods game, which measures monetary contributions, showed that we are intuitively cooperative. As the researchers conclude, "Although the cold logic of self-interest is seductive, our first impulse is to cooperate."[9]

How do we extend this cooperative impulse to other species in order to support their survival? Perhaps we start with recognizing the emotional similarities we share with animals and how these similarities have evolved.

Animal Emotions

We're likely most familiar witnessing emotions in our household pets. We may notice, for instance, a dog's wagging tail as a sign of joy or a cat's seeking solitude during an illness. These behaviors act as clues to a pet's emotional life. But how are these animal emotions expressed in the wild?

s-eyerkaufer

The emotions of animals have been witnessed in dolphins, which have been seen protecting humans from the threat of hammerhead sharks and guiding stranded whales back to the open ocean.

Behavioral ecologist Marc Bekoff reminds us that "emotions have evolved as adaptations in numerous species. They serve as a social glue to bond animals with one another and also catalyze and regulate a wide variety of social encounters among friends and foes."[10] The inner lives and emotions of animals have been witnessed in elephants, for example, who mourn the loss of a fellow elephant by covering it with leaves and visiting the remains for several years. In Kenya scientists have seen elephant mothers assisting their young out of a mud hole, guiding them into the safety of a swamp and helping them break through an electric fence. Elephants have also been seen helping injured elephants by spraying dust on a wound and removing tranquilizing darts from them.[11] All these examples point to elephants' empathic behavior.

Other species that have exhibited their emotions include hungry rats that refused to eat if it meant pulling a lever that would shock their kin; male bluebirds that were seen plucking the feathers of mates discovered with another male; and dolphins protecting humans from the threat of hammerhead sharks and guiding stranded whales back to the open ocean.[12]

Scientists who use anthropomorphic descriptions in their research have often been criticized for deviating from a rigorous science-based approach. However, as technology improves and brain research advances, researchers are getting a better understanding of the emotional lives of animals. As Bekoff points out, "Using anthropomorphic language does not have to discount the animal's point of view. Anthropomorphism allows other animals' behavior and emotions to be accessible to us. Thus, I maintain that we can be *biocentrically anthropomorphic* and do rigorous science.... Even if joy and grief in dogs are not the same as joy and grief in chimpanzees, elephants, or humans, this does not mean that there is no such thing as dog joy, dog grief, chimpanzee joy, or elephant grief."[13] As the scientific debate shifts from whether animals have emotions to what kind or degree of emotions they have, animal activists have moved forward with ensuring the rights of nonhuman animals.

Spearheaded by Steven Wise, the Nonhuman Rights Project, for instance, aims to "change the legal status of appropriate nonhuman animals from mere 'things,' which lack the capacity to possess any legal right, to 'persons,' who possess such fundamental rights as bodily integrity and bodily liberty."[14] The acclaimed documentary film *Unlocking the Cage* traces Wise's battle to use the writ of habeas corpus in New York State to achieve the legal status of personhood for chimpanzees.

Wise and the Nonhuman Rights Project are helping to raise awareness about the rights of all wildlife. In November 2016, an Argentinian court using the writ of habeas corpus ruled that a chimpanzee named Cecilia was a "nonhuman animal person" with the right to be transferred to a Brazilian sanctuary. Wise is pursuing similar lawsuits for the legal rights of elephants, orca whales and other wildlife in various countries. As Wise says, "We are on the cusp of changing the legal relationship between many nonhuman animals and humans."[15]

In addition to the Nonhuman Rights Project's work, neuroscientists are advancing the general public's awareness of consciousness research and the presence of emotions in animals. In 2012 a group of cognitive neuroscientists led by Philip Low declared in the Cambridge Declaration on Consciousness that "The absence of a neocortex does not appear to preclude an organism from experiencing affective states. Convergent evidence indicates that non-human animals have the neuroanatomical, neurochemical, and neurophysiological substrates of conscious states along with the capacity to exhibit intentional behaviors. Consequently, the weight of evidence indicates that humans are not unique in possessing the neurological substrates that generate consciousness."[16]

The Cambridge Declaration on Consciousness describes scientific advancements revealing that, in addition to humans, mammals, birds and other organisms such as octopuses have the capacity for being in "conscious states" and showing "intentional behaviors." Our greater understanding of the neurophysiology of animals has the potential to give them greater protection by affirming their similarities (at the consciousness level) to humans.

The need to protect wildlife is supported by showing our similarities with other species from a legal standpoint by Steven Wise and the Nonhuman Rights Project and from a neuroscientific perspective by Philip Low and the Cambridge Declaration on Consciousness. But in addition to implementing greater legal protections and scientific understanding of animals, how do we devise practical, empathic approaches for coexisting with wildlife? One of the approaches gaining traction involves a new form of conservation, namely, compassionate conservation.

Compassionate Conservation

As scientists discover the cognitive and emotional lives of animals, the field of compassionate conservation is taking root. Perhaps the greatest distinction between conventional conservation and compassionate conservation is that compassionate conservation focuses on individual animals rather than on the species as a whole. When each animal is valued in its own right, a compassionate perspective more easily flourishes. Rather than treating wildlife species as anonymous members of a nondescript population to be managed, each individual is carefully taken into consideration and valued when decisions are made.

The four guiding principles of compassionate conservation are: (1) first do no harm, (2) individuals matter, (3) valuing all wildlife and (4) peaceful coexistence. Marc Bekoff points out that "compassionate conservation establishes compassion as the position driving conservation decision-making, and transparently describes what the costs of achieving 'the greater good' actually represent."[17] The compassionate conservation approach challenges us to shift our perspective in the human/animal conflict that inevitably arises in wildlife management. It does not dictate that the interests of wildlife take precedence over human concerns. Rather it outlines a set of guidelines, rooted in an ethical and humane approach, for devising wildlife management practices. Given that humans have rapidly become the dominant species on Earth, compassionate conservation seeks to create practices that support peaceful coexistence with other species and reduce human/animal conflicts.

P. Jeganathan, Nature Conservation Foundation

To reduce tragic human/elephant encounters in Tamil Nadu, India, a series of early warning systems was developed.

Some of the most innovative management plans for compassionate conservation include projects that promote peaceful coexistence with terrestrial mammals including elephants, leopards and coyotes. The Valparai plateau in Tamil Nadu is home to the second largest concentration of elephants in India (about three herds totaling 100 elephants) and 70,000 residents who live in nearby villages. Over the last century the rainforest has been cleared for tea and coffee plantations, leaving isolated habitat fragments, water sources and paths used by both elephants and people. The competition for food and habitat led to human/elephant encounters that resulted in 39 human deaths between 1994 and 2013.[18]

These tragedies inspired M. Ananda Kumar and his colleagues at the Nature Conservation Foundation to search for a solution that promotes human coexistence with elephants. The multifaceted solution required them to first understand the movement patterns of the elephants and the workers in the limited space. Then they developed a series of early warning systems for the workers including: using the "crawl" bar in local cable television programs to alert villagers to the location of elephants; sending block SMS messages to 4,800 families warning them about the location of elephant sightings; and, for areas that are remote and not covered by TV or text messages, establishing a system of 25 flashing light units that villagers activate to warn their neighbors of elephants seen nearby. This comprehensive early warning system, which has successfully eliminated accidents and deaths from elephant encounters, is being implemented in West Bengal and Kerala.[19] One of the lessons learned from the Valparai initiative is that there are no quick technological fixes. The solutions need to be tailored to each site. In West Bengal, for instance, the elephant awareness program also includes building public toilets so that villagers are discouraged from defecating in the forest, where they are more likely to encounter elephants.

Another example of adapting to peacefully coexist with wildlife is taking place in Mumbai, India, where approximately 35 leopards live among 250,000 residents inside the Sanjay Gandhi National Park, surrounded by 20 million residents in greater Mumbai.[20] For many years, as Mumbai's population has increased, leopards, which rely on stray dogs, livestock, rodents and wild boar to survive inside the 40-square mile urban forest, have had fatal encounters with humans. In fact, a capture and translocation program to remove leopards from afflicted areas resulted in an unexpected increase in attacks from four people a year to 44 over three years. Park scientists learned that this increase was likely due to leopards finding themselves in new and unfamiliar territory with unknown prey, competing with other leopards for limited resources and unexpectedly encountering residents who were uneducated about how to live with leopards as their neighbors. These factors forced leopards to move to areas near residential neighborhoods where livestock and stray dogs were easier prey and where unfortunately human encounters were more likely.[21]

When the leopard translocation program ceased and an education initiative and media awareness program was implemented for residents inside the park and in the surrounding areas, the human/leopard conflicts dropped significantly. Changes such as: controlling the stray dog population, improving lighting, housing and sanitation conditions, proper disposal of garbage and protecting livestock led to a coexistence arrangement that protected the leopards and the people. There was also a shift in the mindset from focusing on "problem animals," which led to undesirable outcomes, to "problem locations," where efforts ensued to find place-based solutions. As T.R. Shankar Raman points out, "Changing the focus from leopard to location also implies a change from reactive measures (such as capture after conflicts occur) to pro-active efforts (such as making safer surroundings to pre-empt attacks) that reduce negative interactions while enabling people and leopards to share the landscape."[22]

In North America, Project Coyote is spearheading the promotion of compassionate conservation and "shifting negative attitudes toward coyotes, wolves and other maligned predators by replacing ignorance and fear with understanding, respect and appreciation."[23] Founder and executive director Camilla Fox believes in the intrinsic value of nonhuman species stating, "We must recognize that animals have value—above and beyond their utility to humans. They have intrinsic worth independent of humans. Moreover, in an increasingly urbanizing world where humans and wild animals are coming into more contact, compassionate conservation gives us the opportunity to reevaluate our relationship to the wild and with the individual animals with whom we interface."[24]

Over 100,000 predators—including coyotes, wolves, bears, mountain lions, foxes and bobcats—are killed annually in the United States by the US Department of Agriculture's Wildlife Services agency. Coyotes alone account for at least half a million deaths—equivalent to one every minute—killed by government agencies and private individuals, costing taxpayers $100 million a year.[25]

Project Coyote's Reforming Predator Management program is shifting the mindset from killing these carnivores to the coexistence of people, livestock and wildlife. In 1996 Project Coyote began working with the Marin County, California, government to stop contracting with the federal government's Wildlife Services, and by 2000 the county had implemented the Marin County Livestock and Wildlife Protection Program to manage ranchers' livestock losses from predators. Rather than relying on the federal government's use of poisons, snares and "denning," where coyote and fox pups are killed in their dens, Project Coyote promotes nonlethal strategies for livestock protection and urban coexistence with coyotes. Nonlethal methods for ranchers include: fencing, fladry (using red nylon flags hanging from a rope on a fence), livestock guardian animals such as dogs, llamas, alpacas and donkeys, night corrals and foxlights, which mimic humans with flashlights.[26]

Since 2005 the Marin County nonlethal livestock protection program has reduced livestock losses from 5.0 to 2.2 percent and brought down costs by $50,000.[27] In addition, Marin agricultural commissioner Stacy Carlsen noted, "This innovative model sets a precedent for meeting a wider compass of community needs and values where both agriculture and protection of wildlife are deemed important by the community. The success of our county model has set the trend for the rest of the nation."[28] In urban settings, programs such as Coyote Friendly Communities educate residents about the ecological value of coyotes and provide training to animal control officers on dealing with coyote/human conflicts.[29]

In Valparai, Mumbai and Marin County, compassionate conservation called for a reevaluation of wildlife. Instead of thinking about wild animals merely as a resource or as a nuisance to be exterminated, biologists and researchers are shifting toward recognizing the intrinsic value of these species, which is not based on how we can benefit from them or what they can do for us. This intrinsic value combines animal rights and beauty, a priceless quality that is one of the strongest motivators for conservation. We value what we appreciate and beauty is at the heart of what we appreciate about nature.

Rewilding Our Hearts

Conservation biologists and environmental activists such as Michael Soulé, Reed Noss, Caroline Frazer and Dave Foreman have been working for decades on rewilding projects aimed at preserving large tracts of land for use by carnivores and other species. As director of the Rewilding Institute, Foreman emphasizes that a successful approach for healthy ecosystems requires three key components: (1) large swaths or cores of protected land for migration and habitation, (2) corridors that connect the different protected fragmented areas since they are not sufficiently big to support the needs of top predators and keystone species and (3) the presence of large predator carnivores such as wolves, cougars, wolverines, jaguars, bears and otters. These three components—cores, corridors and carnivores—are collectively known as the three Cs necessary for rewilding continental-sized landscapes.[30]

Rewilding initiatives are ambitious endeavors requiring decades of work and the coordination of hundreds of organizations. Nevertheless, successful projects are emerging. Over the last 20 years, for example, the Yellowstone to Yukon Conservation Initiative, known as the Y2Y project, is working with 300 partners throughout Canada and the United States to protect a contiguous core habitat. This effort includes: working with legislators to establish new national parks, restoring habitats, building wildlife overpasses, banning oil and gas developments and building wildlife-friendly fencing.[31]

In Europe, the European Green Belt initiative is establishing a protected ecological corridor along the former border between Western and Eastern Europe stretching over 7,500 miles from the Barents Sea in Norway, through the Baltics, Central Europe and the Balkans and ending at the Black Sea. Ironically, the iron curtain that separated Europe for 40 years during the Cold War created wildlife habitats that are currently being protected.[32] In South America, Tompkins Conservation, in partnership with the Chilean government, has established over 10 million acres (an area approximately the size of Switzerland) of new national parklands in Chile and continues its work in wildlife recovery, landscape restoration and ecological agriculture projects in Chile and Argentina.[33] The Y2Y, European Green Belt and Tompkins Conservation projects are grounded on big visions aiming to promote a compassionate view of nature for working with wildlife and human needs.

But how does rewilding nature mirror rewilding ourselves? Perhaps habitat fragmentation mirrors our own disconnection from nature and the way to heal that dysfunction is by cultivating our compassion toward other species. Instead of attempting to heal that disconnect by living in isolation from modern culture, the solution lies in strengthening our empathic values and sharing the landscape with wildlife. As Bekoff reminds us, "Rewilding our hearts doesn't mean becoming an 'off-the-grid' survivalist, a radical 'back-to-the-land' activist, or a hard-core outdoorsperson. It means, simply, acting with compassion and love for nonhuman animals and for the natural world that is our shared home."[34] Reestablishing our connection with nonhuman animals and nature as a whole not only rewards us with feeling connected to something larger than ourselves but also acts as a catalyst for unveiling our creativity, imagination and courage.

Compassion is a bridge linking our inner selves with the natural world. Nurturing our coexistence with other people and with animals begins with seeing what we have to offer to others. As Mary Reynolds Thompson asks, "But what if the process of rewilding the Earth began with rewilding our souls? If we truly grasp the interconnectedness between all living things, doesn't it follow that every change within us will be reflected in the whole? If we reroot ourselves in the rhythms, wisdom, and patterns that created this planet and our own flesh and feelings, what might be possible for the Earth and all her inhabitants? What if healing the world really does start from within?"[35]

Rewilding the Earth begins with developing a compassionate approach towards all species and making ecosystems whole again. The three Cs—cores, corridors and carnivores—needed to rewild the Earth have a parallel set of three Cs for rewilding and healing ourselves, namely being conscious, caring and courageous. By being conscious we tap into our true nature and act from a grounded and authentic place. Being caring reveals how we can contribute to our community. Being courageous helps us go beyond any limiting beliefs to achieve lasting solutions. Together these qualities define an ethic for living in harmony with the natural world.

QUESTIONS AND ACTIVITIES

- What childhood experiences shaped your values and attitudes about animals and nature?

- How are you and/or your community enhancing habitats that nurture wildlife in your home town?

- Have you had a human/animal conflict in your community? How was it resolved?

- In what ways have you nurtured compassion in your life?

- What role does nature play in giving you insights and wisdom?

- How does new research about the consciousness and sentience of nonhuman animals change your views of other species?

- In what ways can you improve your coexistence with animals in your neighborhood?

8 An Ecocentric Ethic

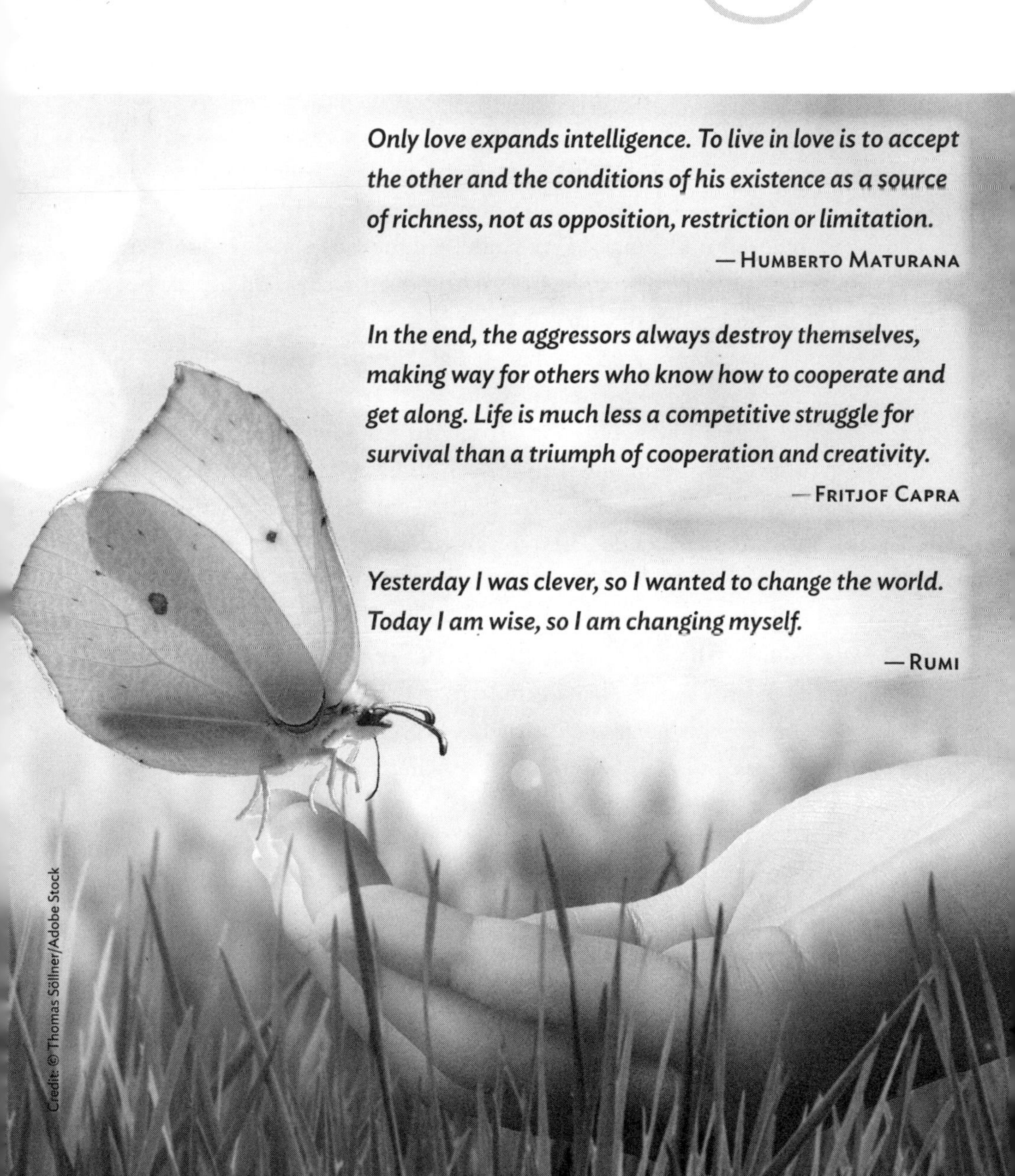

Only love expands intelligence. To live in love is to accept the other and the conditions of his existence as a source of richness, not as opposition, restriction or limitation.

— Humberto Maturana

In the end, the aggressors always destroy themselves, making way for others who know how to cooperate and get along. Life is much less a competitive struggle for survival than a triumph of cooperation and creativity.

— Fritjof Capra

Yesterday I was clever, so I wanted to change the world. Today I am wise, so I am changing myself.

— Rumi

Credit: © Thomas Söllner/Adobe Stock

One of my most bizarre camping experiences happened while I was working with a botanist to identify and collect rare and endangered plants in the south of Chile. On one of our longest days, when we had been searching for a campsite for hours, we found a promising dirt road, which we drove down until it dead-ended. There, in the dark through the headlights, to our delight we saw a clearing. We were thrilled to finally find a place to sleep.

While getting our tent ready we noticed a peculiar foul smell—a combination of rotten eggs and rotten food—but we ignored it as much as we could. I noticed the soil was soft and spongy and my first thought was that we must be on a peat bog. When we finally got into our sleeping bags, the foul smell lingered, but at that point we were exhausted and just fell asleep.

The next morning we were awakened at dawn by the sound of trucks and bulldozers. When I opened the tent flap I saw several garbage trucks dumping their loads near us and bulldozers scooping and moving the waste. We had pitched our tent in a landfill! There were construction waste, sewer pipes and garbage all around us.

This experience reminds me of the expression, "If it looks like a duck, swims like a duck and quacks like a duck, then it probably is a duck." The clues about the landfill were all around us, yet we preferred to dismiss them. Similarly, the clues about the benefits of being in nature (such as improved health, enhanced creativity, greater happiness and well-being) are everywhere, yet many of us are not embracing them.

An Integral Perspective

It's time to reimagine our relationship with nature from the inside out, incorporating emotional, social and biological perspectives. The emotional perspective calls us to cultivate self-care by tuning in to our body, emotions, mind and spirit. By paying close attention to how we feel in a natural setting, we nourish our care for nature. We can be transformed when our senses experience the sights, sounds, smells, tastes and textures of nature. Such experiences can help us become more caring about other species and the health of their habitats. When we recognize that we are not alone and that we share with other species what sustains us all, such as air, water and food, we see our impact on nature from a new, holistic perspective that leads us to take action to alleviate the suffering of others.

What if rather than intellectualizing the plants, animals and processes in nature we just experience them, much as children do when they explore the outdoors motivated by their curiosity and the mystery of life? Rachel Carson, one of the pioneers of the modern environmental movement, noted how as adults we can support children to feel and experience for themselves the impact of nature rather than teaching them about it. As she points out, "I sincerely believe that for the child, and for the parent seeking to guide him, it is not half so important to *know* as to *feel*. If facts are the seeds that later produce knowledge and wisdom, then the emotions and the impressions of the senses are the fertile soil in which the seeds must grow."[1] The fertile soil that nurtures our emotional bond with nature as children (and later as adults) will germinate the seeds that turn into our life-long love for nature.

From a social perspective, we see that at our core we are social beings with a range of emotions, desires and motivations. As the world population grows and our cities rapidly expand, the equitable distribution of food, energy, education and healthcare is becoming paramount. The solutions to these pressing challenges require a collaborative approach that recognizes that in order for me to thrive my neighbor must thrive as well.

Therefore, when we see environmental damage done by others—such as government institutions, corporations or individuals—we need to evaluate all the factors comprehensively. At this level, it's not primarily about my well-being but more about the greater good. Tenzin Gyatso, the 14th Dalai Lama, calls for humanity to develop a sense of "universal responsibility" where we honor the needs of others because we recognize that we are interconnected. As he says, "We consider our own interests as the most important ones and disregard others' interests. So that creates a problem. In reality our interests and others' interests are very much interconnected.... [O]ur centuries-old concept that "us" and "them" are independent is, I think, outdated. Now, particularly in these modern times, with the economic conditions, the environmental issues and the sheer size of the population, everything is interdependent."[2]

The biological perspective expands the view that we are interconnected to encompass our interdependence with the natural world. To find viable solutions to the complex problems we are confronting such as climate change, pollution and loss of biodiversity, we need to work together using a systems approach. This approach sees what systems theorist Fritjof Capra describes as the "connectedness, relationships, patterns, and context" in nature. "According to the systems view the essential properties of an organism, or living system, are properties of the whole, which none of the parts have. They arise from the interactions and relationships between the parts."[3] Instead of focusing on the individual parts of an ecosystem, such as the soil, plants or species, we view the connections and relationships among these parts. The same applies to human-built systems, such as business, healthcare and education.

Five Global Trends

The new perspective will help us deal with five global trends we're facing: (1) the migration of people from rural areas to cities; (2) people living with wildlife in cities and in urban-rural fringe areas; (3) the reduced time children and adults are spending outdoors in nature; (4) the fewer nature references in popular culture; and (5) the advances in cognitive science showing the benefits of being in nature.

Rural to urban migration is creating strains in available resources, especially in cities with over 10 million residents. Over half the world's population lives in cities, which have become incubators for numerous innovative solutions in urban agriculture, renewable energy and transportation systems. The global green cities movement is inspiring community leaders to proudly "green" their cities and showcase their success stories to the world. Urban planners are designing and implementing innovative urban parks and wildlife corridors, recognizing their health benefits for city residents as well as for wild species.

As the human population grows, living with wildlife in urban areas has become inevitable. Michael Rozenweig of the University of Arizona calls the conservation of wildlife in urban areas reconciliation ecology, defining it as "the science of inventing, establishing and maintaining new habitats to conserve species diversity in places where people live, work and play."[4] Rozenweig recognizes that this strategy will not work for all species, especially mammals that need large tracts of land, but for many species, such as butterflies, bats and foxes, restoring and creating natural habitats in dense urban areas may be a viable way to help them flourish, as has already been proven in several cities. Strategies for coexisting with wildlife in urban-rural fringe areas involve education about managing human-wildlife encounters and adapting to wildlife living there.

The reduced time that children and adults spend in nature, resulting in what Richard Louv calls "nature deficit disorder," (see chapter 3) is changing how we perceive and experience the natural world. While the full implications of this evolving shift remain unknown, if children experience nature as an intellectual concept viewed through a screen the number of people with an emotional connection to nature may be reduced.

Because childhood experiences such as playing in the woods or on the beach have an important role in developing our love of nature, what will happen if these experiences are absent? Since technology is everywhere and not likely to go away, perhaps what matters is how we use it. We can, for instance, use apps that help us learn about constellations and identify plants and birds, but we still need to experience delight in the awe and beauty of unexpected events that often occur in nature when we quietly listen and observe what's going on around us. But as Rachel Carson reminded us "it is not half so important to *know* as to *feel*."[5]

Andrés Edwards

The way nature operates without creating waste,
harnessing energy from the sun through photosynthesis and recycling
water and nutrients, serves as inspiration.

In modern culture we not only spend less time outdoors in nature, but there are also fewer nature-related words in fiction, movies and popular songs. Social scientists Selin Kesebir and Pelin Kesebir selected 186 nature-related words used to identify nature in general (lake, autumn...), flowers (rose, foxglove...), trees (cedar, willow...) and birds (hummingbird, finch...). They discovered that "For every three nature-related words in popular songs of the 1950s, for example, there is slightly more than one 50 years later." In 1957, for instance, the hit songs included "Honeycomb," "Butterfly" and "Dark Moon," among others.[6]

The loss of nature-related words is also impacting children's vocabulary. Recently over 50,000 people signed a petition asking the *Oxford Junior English Dictionary* to reinstate nature-related words that it had removed from its current edition.[7] Words like: "acorn, ash, beech, bluebell, buttercup, catkin, cowslip, cygnet, dandelion, fern, heron, ivy, kingfisher, lark, mistletoe, nectar, newt, otter, pasture and willow" were removed and words including: "attachment, block-graph, blog, broadband, bullet-point, celebrity, chatroom, committee, cut-and-paste, MP3 player and voice-mail" were added.[8] The result of this emphasis on technology-oriented words is a culture where children can identify more Pokémon characters than plants and animals.

Finally, the current explosion of knowledge in cognitive science shows that we are happier, healthier, kinder, more creative and more grateful when we spend time in nature (see chapters 3, 6 and 7). Perhaps in the not too distant future medical prescriptions will read: "30 minutes of nature twice a day for six months." What a bargain the nature prescription would be compared to the medical costs of treating stress, high blood pressure and depression.

How does the knowledge of the benefits of being in nature lead us to a more compassionate approach toward people and nonhuman species? If nature helps us heal, how do we translate that into living in balance with the natural world? The answer may lie in nurturing an ecocentric ethic.

Reciprocity Through Nature

An ecocentric ethic involves a reverence for life and the surrounding natural environment. This perspective evaluates our actions based on whether they support life on Earth. Biologist Janine Benyus says, "Life creates conditions conducive to life"[9]—that is, life is constantly evolving and regenerating within the limits of living systems. We are a part of this regenerative system, which has been evolving for millennia. An ecocentric ethic focuses on the critical role of our interdependence with the natural world. The way in which nature operates without creating waste, harnessing energy from the sun through photosynthesis and recycling water and nutrients serves as inspiration.

How can we play our roles while honoring and enriching other species? How do we align with nature's living rhythms? Joseph Campbell reminds us of the importance of synchronizing with the beat of the universe, stating, "The goal of life is to make your heartbeat match the beat of the universe, to match your nature with Nature."[10] Matching our heartbeat with the beat of the universe implies a rhythmic exchange where we respectfully take and give back. We synchronize the way we live with the way other species live, meeting our needs as well as theirs.

The balanced exchange of an ecocentric ethic comprises three aspects, the ABCs of honoring life on Earth: (1) awareness, (2) belonging and (3) caring. The foundation of these important qualities is a sense of reciprocity, recognizing that as interdependent species we have a moral obligation to give back as well as take. Douglas Christie states in *The Blue Sapphire of the Mind* "that our ecological commitments, if they are to reach mature and sustainable expression, need to be grounded in a sense of deep reciprocity with the living world. And that this sense of reciprocity must be cultivated over time, in a process of deepening awareness and growing ethical maturity...."[11]

Our ethical maturity is rapidly evolving as cognitive science studies show the benefits of being in nature, and with the interest in reducing stress by tuning in to our bodies through meditation, yoga and other contemplative practices. One of the organizations exploring the benefits of awareness is the HeartMath Institute. Founder Doc Childre is committed to providing "tools that connect us with 'the heart of who we truly are.'" Using a science-based approach, HeartMath programs are helping people to "bring their physical, mental and emotional systems into balanced alignment with their heart's intuitive guidance."[12] We're familiar with the expressions "follow your heart," "listen to your heart" and "speak from your heart." The HeartMath researchers are delving deeply into measuring what goes on emotionally and physiologically when our hearts and minds are in alignment and how we can promote this balance.

Childre and his colleagues are expanding their work beyond the individual to activate the "heart of humanity" by exploring the potential of awakening our mental, emotional and spiritual capacities at a global scale. One of the HeartMath Institute's large-scale programs is the Global Coherence Initiative, which looks at the relationship between the Earth's magnetic field and our health and behavior by using a network of magnetometers gathering data at different locations around the globe. The researchers hope to highlight heart-based connections that will lead to intuitive solutions to global problems. As Childre points out, "As more of humanity practices heart-based living it will qualify [as] the 'rite of passage' into the next level of consciousness. Using our heart's intuitive guidance will become common sense—based on practical intelligence."[13]

Scientists continue to advance the "rite of passage" into a higher level of awareness through numerous experiments revealing remarkable discoveries. These discoveries point to a new way of understanding how the energy in our bodies is connected to the broader energy field in the world. As Lynne McTaggart describes in *The Field*, "At our most elemental, we are not a chemical reaction, but an energetic charge. Human beings and all living things are a coalescence of energy in a field of energy connected to every other thing in the world. This pulsating energy field is the central engine of our being and our consciousness, the alpha and the omega of our existence."[14] Our

Ian McDonnell

Neuroconservation studies our neurological responses to nature, especially water, as a conservation strategy.

central engine, namely our heart's intuitive guidance, helps us navigate our life's journey and make the wisest decisions when it's in sync with nature's pulsating energy field.

But rather than making decisions in isolation, we benefit from nurturing a sense of belonging, the second aspect of an ecocentric ethic. When we feel that we belong to something bigger than ourselves—to life on the planet and its energy field—we are empowered as active stewards rather than idle bystanders. Philosopher Marshall McLuhan said, "There are no passengers on Spaceship Earth. We are all crew."[15] And activist Starhawk describes her "transformation from a tourist in nature to an inhabitant,"[16] emphasizing the shift from observer to participant in the constantly evolving natural world. But what underlies that sense of belonging to nature and by extension to the universe? Marine biologist Wallace J. Nichols believes that neuroscience may reveal some answers.

Nichols' work emerged from his early days studying sea turtles in Mexico and realizing that beyond scientific facts, understanding the role of our emotional connection to nature is essential for the successful implementation of conservation strategies. As Nichols recounts, "My most successful work as a biologist with sea turtles was based on that emotional connection. From working with turtle hunters that are now turtle protectors to working with tourists to working with scientists, it is all based in that emotional connection that we aren't supposed to talk about."[17]

In raising awareness about the connections between our emotions and nature, Nichols is one of the pioneers in a new field he calls neuroconservation, which studies our neurological responses to nature as a conservation strategy. With a particular interest in how our brain responds to being in and around water, he believes that a key lies in our emotional connection to the oceans, which blends emotional and scientific inquiries. As he points out, "It's time to drop the old notions of separation between emotion and science.... Emotion is science.... It's likely, maybe even certain, that the greatest unexplored mysteries of the sea are buried not under a blanket of blue, but deep in the human mind."[18]

Nichols' book *Blue Mind* brings attention to the benefits of being in nature, particularly the sea, and calls for further research into the role water plays in reducing stress and making us happier and more creative, compassionate and grateful to each other and the natural world. To that end, Nichols' Blue Marbles Project aims to distribute a blue marble, symbolizing the Earth, to every person on the planet, an ambitious objective that would have each of us hand a blue marble as a token of gratitude to someone else, who would then pass it on, creating a "low-tech, slow-motion global art project and a clear reminder that everything we do on this water planet matters."[19] The project has gained visibility as celebrities including Jane Goodall, E.O. Wilson and the Dalai Lama have received blue marbles and helped to bring awareness to our sense of belonging to the water planet we all inhabit. Nichols is also spreading the global conversation about our emotional connection to water through the Blue Mind conference, which brings together an eclectic group of scientists, activists, lawyers, musicians and artists, among others, to share ideas and advance research into the effects of what Nichols calls "our brains on water."

The third aspect of an ecocentric ethic is caring. Caring has an internal component, namely, self-care, and an external one, care for the Earth. Self-care includes the many ways we keep our mind, body, emotions and spirit healthy, such as by eating healthful food, exercising and participating in social activities, creative endeavors and professional development. The outward focus involves caring about the Earth's biosphere and its living systems on which we depend for our survival such as clean air and water, healthy soil, biodiversity and a stable climate.

To care for the Earth we must know it, and to know it we must nurture our compassion and love for all life. As Wendell Berry reminds us, "We have the world to live in on the condition that we will take good care of it. And to take good care of it we have to know it. And to know it and to be willing to take care of it, we have to love it."[20] Our compassion, care and love for nature blossom when we nurture our emotional connection to all living beings.

Flourishing with Nature

The ecocentric ethic gives us a useful compass for how to live in harmony with the Earth's living systems. Often the catalyst that sparks these life-affirming values is an experience we had, sometimes at a young age, that forever changed how we feel about nature and in many cases directed our life's journey. In *A Sand County Almanac,* for instance, conservationist Aldo Leopold wrote about the emotional impact of his encounter with a dying wolf, which turned him toward his life-long conservation work:

> We reached the old wolf in time to watch a fierce green fire dying in her eyes. I realized then, and have known ever since, that there was something new to me in those eyes—something known only to her and to the mountain. I was young then, and full of trigger-itch; I thought that because fewer wolves meant more deer, that no wolves would mean hunters' paradise. But after seeing the green fire die, I sensed that neither the wolf nor the mountain agreed with such a view.[21]

Similarly, environmental activist Paul Watson recounts his experience while working with Greenpeace shielding a pod of whales from being harpooned. Unexpectedly, an injured sperm whale "with an eye about the size of a dinner plate" swam swiftly toward him, passing within 10 feet. Then the whale looked up and everything changed "when a wounded Sperm Whale could've killed me and chose not to do so."[22]

Both Watson's and Leopold's experiences sparked their commitment to protecting wildlife and propelled them into careers in environmental conservation. For many of us, with perhaps less dramatic encounters, our love of nature grew from our appreciation for the beauty and delight we experienced while exploring the natural world, from observing the patterns of a butterfly's wings or the crashing waves at the beach or from the joy of swimming in a lake on a summer's day. Whatever the experience, there was an underlying emotional aspect, something that fascinated us or filled us with awe and evolved into a deeper love of the natural world.

A Legacy of Regenerative Coexistence

In the Lakota Sioux language nature encompasses "all our relations," highlighting our kinship with all plants, animals and the Earth. Beyond what we "take" from nature for our well-being, what can we give back? How do we show our gratitude? Perhaps the first step is to develop a legacy of regenerative coexistence, emulating the regenerative aspects of nature such as the ways nature renews and replenishes the Earth through the microorganisms that nourish the soil and purify the water, the plants that produce oxygen through photosynthesis and the bountiful fruits, nuts and seeds produced every season.

Perhaps the seeds for our transformation begin with cultivating gratitude, beauty and compassion. We can embrace these qualities and use them as markers, or "cairns," on our journey toward regenerative coexistence. We can be grateful for the work that our friends and colleagues are doing to improve the lives of people and the planet and for the Earth that sustains us. We can celebrate and protect the beauty that surrounds us in nature as well as the beautiful acts of kindness and generosity that people share every day. We can generate more compassionate attitudes to "all our relations," including our neighbors and other species that we may never encounter, by taking actions for their well-being.

Gratitude, beauty and compassion become the ingredients for nurturing our love for all life. Although it's one of the most powerful universal emotions, love is rarely recognized as a motivating force in science. In fact, it's often dismissed as antithetical to scientific research but it's actually a powerful force in science that helps us care for and protect nature. As environmental educator David Orr points out, "Science, at its best, is driven by passion and emotion. We have emotions for the same reason we have arms and legs: They have proved to be useful over evolutionary time. The point in either case is not to cut off various appendages and qualities, but rather to learn to coordinate and discipline them to good use."[23]

The love that drives us to care for nature will lead us to stand strong and confront the greed and shortsightedness that often accompany the economic forces that destroy natural habitats. In the face of mounting threats to nature, the question that arises is: what can I do? The answer lies in discovering our own intuitive compass and aligning it with the pulse of the planet.

Doing What We Can

The daunting environmental challenges we face often lead us to feelings of grief and despair, but aligning with nature yields positive emotions that help us feel alive. Understanding our reciprocal connection with the natural world gives us the clarity to take actions that are appropriate for us and for nature.

The First Nations people of the northern Pacific Rim region tell the story of a hummingbird doing what it can to protect its home:

> A fire had begun in the forest and was in danger of raging out of control. Terrified, many of the animals fled before it took over. But the hummingbird flew to the nearest water, collected a droplet and flew back to the fire dropping the water onto it. As she flew back and forth to protect her habitat, first the bear, and then the owl, the snake and cougar each called out: "Bird, what are you doing?" The hummingbird answered them all in the same way: "I am doing what I can."[24]

May nature inspire us with her wisdom to do what we can to protect our home, the Earth, by collecting infinite droplets of gratitude, beauty, compassion and love to grow a regenerative world for all.

QUESTIONS AND ACTIVITIES

- Have you experienced being in alignment with nature's living rhythms? How did it feel?

- When was the first time you remember having an emotional connection with nature?

- What does using your heart's intuitive guidance mean to you?
- How do you nourish your ABCs: awareness, belonging and caring?
- What is the role of regenerative coexistence in your life?
- What contemplative practices help you stay grounded and vibrant?
- How do you integrate gratitude, beauty, compassion and love into your life?

Resources

(in their own words...)

Chapter 1: Aligning with Nature

Mutual Flourishing for Self and Earth: Chara Armon, PhD
mutualflourishing.org

My work supports healing and flourishing for planet and people by inspiring you to listen to your heart's compassion for all life, your soul's desire to serve, and your body's yearning to live in harmony with the Earth and all her inhabitants. I support women who seek to contribute to healing the human-Earth relationship and nurturing human wellness on a healthy planet, while also nurturing themselves.

Center for World Indigenous Studies — cwis.org

We are a global community of activist scholars advancing the rights of indigenous peoples through the application of traditional knowledge.

Cultural Survival — culturalsurvival.org

Our work is predicated on the United Nations Declaration on the Rights of Indigenous Peoples. We engage opportunities to leverage our experience and leadership in advocacy, media, public education, programs, and in providing platforms to amplify and empower the voices of Indigenous Peoples as they work to claim their rights to self-determination, their lands, cultures, and precious ecosystems that are essential to the whole planet.

Ecology Global Network — ecology.com

The ECOLOGY Global Network™ Mission Statement: To use the modern tools of information and communication to inform, educate and inspire the global community to respect, restore and protect our natural and human world, and to encourage all people to become stewards of the environment in which we live.

Food Timeline foodtimeline.org

Ever wonder how the ancient Romans fed their armies? What the pioneers cooked along the Oregon Trail? Who invented the potato chip...and why? So do we!!! Food history presents a fascinating buffet of popular lore and contradictory facts. Some experts say it's impossible to express this topic in exact timeline format. They are correct. Most foods are not invented; they evolve. We make food history fun.

Discover John Muir discoverjohnmuir.com

Use this website for: information sources; activities that help connect with Muir's messages, stories and adventures; Muir-related events; examples of how people engage with Muir's ethos in their studies, work and leisure time as well as through the John Muir Award.

John Muir Trust johnmuirtrust.org

Our mission: To conserve and protect wild places with their indigenous animals, plants and soils for the benefit of present and future generations. Our vision: A world where wild places are protected, enhanced and valued by and for everyone.

The Nature Institute natureinstitute.org

Mission: At The Nature Institute, we develop new qualitative and holistic approaches to seeing and understanding nature and technology. Through research, publications, and educational programs we work to create a new paradigm that embraces nature's wisdom in shaping a sustainable and healthy future.

Resurgence resurgence.org

The Resurgence Trust is a UK-based educational charity that has become the flagship voice of the environmental movement across the globe. Our mission is to inspire each other to help make a difference and find positive solutions to the global challenges we now face.

RSF Social Finance rsfsocialfinance.org

Mission: To create financial relationships that are direct, transparent, personal and focused on long-term social, economic and ecological benefit. We achieve our mission by: Offering investors and donors dynamic ways to align their money with their values; Connecting entrepreneurs with diverse forms of capital; Empowering leaders and advancing innovations in the field of social finance.

Chapter 2: Awe and Beauty

Big Think bigthink.com

Big Think is a knowledge forum. In our digital age, we're drowning in information. The web offers us infinite data points—news stories, tweets, wikis, status updates, etc.—but very little to connect the dots or illuminate the larger patterns linking them together. Here at Big Think, we believe that success in the future is all about knowing the ideas that allow you to manage and master this universe of information. Therefore, we aim to help you move above and beyond random information, toward real knowledge, offering big ideas from fields outside your own that you can apply toward the questions and challenges in your own life.

Center for Humans and Nature humansandnature.org

We're a group of engaged and curious thinkers who understand that ideas matter. The Center for Humans and Nature partners with some of the brightest minds to explore human responsibilities to each other and the more-than-human world. We bring together philosophers, ecologists, artists, political scientists, anthropologists, poets and economists, among others, to think creatively about a resilient future for the whole community of life.

Ralph Waldo Emerson emersoncentral.com

List of Writings by Ralph Waldo Emerson. The following are the texts available on this web site. For additional sites with Emerson materials please see the Ralph Waldo Emerson Home Page or The Transcendentalists web site.

Futurity futurity.org

Futurity features the latest discoveries by scientists at top research universities in the US, UK, Canada, Europe, Asia, and Australia. The nonprofit site, which launched in 2009, is supported solely by its university partners (listed below) in an effort to share research news directly with the public.

Greater Good Magazine greatergood.berkeley.edu

Greater Good magazine turns scientific research into stories, tips, and tools for a happier life and a more compassionate society. Through articles, videos, quizzes, and podcasts, we bridge the gap between scientific journals and people's daily lives, particularly for parents, educators, business leaders, and health care professionals. *Greater Good* magazine is published by the Greater Good Science Center (GGSC) at the University of California, Berkeley. Since 2001, the GGSC has been at the fore of a new scientific movement to explore the roots of happy and compassionate individuals, strong social bonds, and altruistic behavior—the science of a meaningful life.

Harvard University Sustainability **green.harvard.edu**

Harvard University is devoted to excellence in teaching, learning, and research, and to developing leaders in many disciplines making differences globally. While Harvard's primary role is to address global challenges, such as climate change and sustainability, through research and teaching, the University is also focused on translating research into action. Harvard is using its campus as a living laboratory for piloting and implementing solutions that create a sustainable and resilient community focused on health and well-being.

Heal Naturally **realnatural.org**

Heal Naturally is published by Realnatural, Inc., a company dedicated to the health of our bodies and our planet. The Heal Naturally site offers the latest in cutting edge natural health research and information, and original articles. We seek to educate by virtue of scientific research and clinical evidence about those traditional remedies our ancestors used for centuries to maintain health and vitality without ruining the environment and our bodies.

Aldo Leopold Foundation **aldoleopold.org**

Mission: The Aldo Leopold Foundation's mission is to foster a land ethic through the legacy of Aldo Leopold. Vision: Our vision is to weave a land ethic into the fabric of our society; to advance the understanding, stewardship and restoration of land health; and to cultivate leadership for conservation.

The Mystica **themystica.com**

The Mystica is an ancient Online Encyclopedia of knowledge. It contains information about ancient Mythology (African, Roman, Greek, Asian, Greco Roman, Egyptian, Buddha, Hindu, Native American and South American), Folklore, Ancient, Past and Present Beliefs, Paranormal, Occultism, Religion, Dreams and general knowledge. The objective of the Mystica is to objectively describe ancient, past and present ideas, definitions, concepts, beliefs, and practices in the worlds of the occult, mysticism, paranormal and general knowledge. The information in the Encyclopedia is organized in the following sections. Each section contains articles that cover the topics with definitions, ideas, related articles and also related external sources.

The Online Gadfly gadfly.igc.org/index.htm

The Online Gadfly is the website of "Gadfly Enterprises," administered by Ernest Partridge (hereafter, "The Gadfly"), a consultant, writer and lecturer in the field of Environmental and Applied Ethics. This website contains original essays, an assortment of papers (published, unpublished, and in progress), news and notes, bibliographies and other research tools, and much more, most of these pertaining to issues in environmental ethics, politics and public policy. The Online Gadfly will also serve as a "gateway" to other web sites, publications, organizations and educational institutions devoted to progressive political ideals and to the study of environmental ethics and policy.

Psychology Today psychologytoday.com/us

Psychology Today is devoted exclusively to everybody's favorite subject: Ourselves. On this site, we have gathered a group of renowned psychologists, academics, psychiatrists and writers to contribute their thoughts and ideas on what makes us tick. We're a live stream of what's happening in 'psychology today'. Our magazine, first launched in 1967, continues to thrive.

Jason Silva—Shots of Awe thisisjasonsilva.com

Jason Silva is an Emmy-nominated and world renowned TV personality, storyteller, filmmaker, and sought-after keynote speaker and futurist. Jason is known for hosting 5 seasons of the Emmy-nominated, global hit TV series Brain Games, on the National Geographic Channel, broadcast in over 171 countries. His inspirational videos, Shots of Awe, have received over 100 million views across social platforms. The videos explore topics such as futurism, technology, creativity, the science of awe, disruptive innovation, relationships and mental health.

Smithsonian TweenTribune tweentribune.com

TweenTribune, A Free K-12 Resource for Teachers. Join more than 200,000 registered teachers who are already putting these free Smithsonian Teacher tools to use in their classrooms.

Stanford Encyclopedia of Philosophy plato.stanford.edu/index.html

Welcome to the Stanford Encyclopedia of Philosophy (SEP). From its inception, the SEP was designed so that each entry is maintained and kept up-to-date by an expert or group of experts in the field. All entries and substantive updates are refereed by the members of a distinguished Editorial Board before they are made public. Consequently, our dynamic reference work maintains academic standards while evolving and adapting in response to new research.

Thoreau Society thoreausociety.org

Mission: The Thoreau Society exists to stimulate interest in and foster education about Thoreau's life, works, legacy and his place in his world and in ours, challenging all to live a deliberate, considered life. Vision: The Thoreau Society keeps Thoreau's writings and ideas alive around the globe and across generations.

Tompkins Conservation tompkinsconservation.org/home.htm

We believe that humans have an ethical obligation to share the planet with other species, and that we must reorient our values and activities so that all forms of life can flourish. Toward this end, we direct our energies to park creation, activism, restoration, and ecological agriculture. Throughout diverse programs, we uphold our commitment to a common set of ideals: ecologically grounded local economies; local, renewable energy production; thoughtful, place-appropriate architecture and design; and meaningful work for individuals and communities.

Chapter 3: Health and Well-Being

30x30 Nature Challenge

davidsuzuki.org/take-action/act-locally/30x30-nature-challenge/

What is the 30x30 Nature Challenge? Since 2012, the David Suzuki Foundation has invited Canadians and people around the world to join the 30×30 Nature Challenge. Tens of thousands of people have been inspired to spend 30 minutes outside each day for 30 days. Join us in cultivating the nature habit!

American Horticultural Therapy Association ahta.org

Mission: AHTA is a non-profit membership-driven organization whose mission is to promote and advance the profession of Horticultural Therapy as a therapeutic intervention and rehabilitative modality through: The dissemination of information relating to the principles and practices of horticultural therapy as a treatment modality. The encouragement of professional growth of horticultural therapy practitioners. The establishment of professional standards and a credentialing process for horticultural therapy practitioners. The promotion of research related to the impact of horticultural therapy as a treatment modality. The advocacy of horticultural therapy as a treatment modality to the public, the healthcare industry, the academic community, and the allied professions.

Another Escape anotherescape.com
Another Escape is an outdoor lifestyle, creative culture and sustainable living publication that explores the stories of passionate people, alluring landscapes and intriguing ideas. We cover a disparate selection of subject matter all deep-rooted in exploration, creativity, innovation and discovery, and aim to be a source of inspiration for those who seek an active and considered lifestyle by encouraging an optimistic, forward-thinking and responsible mindset.

Children & Nature Network childrenandnature.org
Vision: A world in which all children play, learn and grow with nature in their everyday lives. Mission: The Children & Nature Network is leading the movement to connect all children, their families and communities to nature through innovative ideas, evidence-based resources and tools, broad-based collaboration and support of grassroots leadership.

Green Cities: Good Health depts.washington.edu/hhwb/
This web site provides an overview of the scientific evidence of human health and well-being benefits provided by urban forestry and urban greening.

International Community for Ecopsychology ecopsychology.org
The International Community for Ecopsychology (ICE) is an informal, international, interdisciplinary virtual community devoted to collective reflection on the questions which arise from an ecopsychological viewpoint. With our website, journal, and blogs we hope to facilitate communications among people with an interest in ecopsychology and related topics.

International Journal of Wellbeing internationaljournalofwellbeing.org
The *International Journal of Wellbeing* welcomes timely original high-quality scholarly articles of appropriate length on the topic of wellbeing, broadly construed. Although focused on original ideas, the *International Journal of Wellbeing* also publishes competent and timely review articles and critical notices. Book reviews are at the request of the editors only.

National Center for Biotechnology Information ncbi.nlm.nih.gov
The National Center for Biotechnology Information advances science and health by providing access to biomedical and genomic information.

National Wildlife Federation nwf.org
Mission and Strategic Plan: We believe America's experience with cherished landscapes and wildlife has helped define and shape our national character and

identity for generations. Protecting these natural resources is a cause that has long united Americans from all walks of life and political stripes. To hunters, anglers, hikers, birders, wildlife watchers, boaters, climbers, campers, cyclists, gardeners, farmers, forest stewards, and other outdoor enthusiasts, this conservation ethic represents a sacred duty and obligation to protect and build upon our conservation heritage for the sake of wildlife, ourselves, our neighbors, and—most of all—for future generations.

Positive Psychology Program positivepsychologyprogram.com

The Positive Psychology Toolkit© is a science-based, online platform containing 190+ exercises, activities, interventions, questionnaires and assessments. It's designed for people who are passionate about helping others improve their lives in meaningful ways, by applying positive psychology. That is why—in collaboration with the world's top universities, organizations, and researchers—we are continuously looking for the latest scientific and practical tools to add to the toolkit.

The Pursuit of Happiness pursuit-of-happiness.org

What we do: Provide science-based information on the life skills and habits needed to enhance well-being, build resilience against depression and anxiety, and pursue a meaningful life. Draw attention to the remarkable links between ancient wisdom and the new science of happiness. Promote the pursuit of happiness through educational programs designed around the resulting knowledge base. At present, we collaborate with educational authorities in the US, China, and South America to integrate the teaching of life skills related to psychological well-being with current curricula.

The Royal Society royalsociety.org

Our Mission and Priorities: The Society's fundamental purpose, reflected in its founding Charters of the 1660s, is to recognise, promote, and support excellence in science and to encourage the development and use of science for the benefit of humanity. The Society has played a part in some of the most fundamental, significant, and life-changing discoveries in scientific history and Royal Society scientists continue to make outstanding contributions to science in many research areas.

ScienceDaily sciencedaily.com

ScienceDaily is one of the Internet's most popular science news web sites. Since starting in 1995, the award-winning site has earned the loyalty of students, researchers, healthcare professionals, government agencies, educators and the general public around the world. With roughly 5 million monthly visitors worldwide, ScienceDaily reaches a global audience.

ScienceDirect sciencedirect.com

What is ScienceDirect? Elsevier's leading platform of peer-reviewed scholarly literature. University libraries and institutions offer ScienceDirect access to their communities of researchers. Researchers, teachers, students, healthcare and information professionals use ScienceDirect to improve the way they search, discover, read, understand and share scholarly research.

Science Magazine sciencemag.org

Science is a leading outlet for scientific news, commentary, and cutting-edge research. Through its print and online incarnations, *Science* reaches an estimated worldwide readership of more than one million. *Science*'s authorship is global too, and its articles consistently rank among the world's most cited research.

VIA Institute on Character viacharacter.org/www/

Our Mission: The VIA Institute on Character is a non-profit organization, based in Cincinnati, Ohio, dedicated to bringing the science of character strengths to the world through supporting research, creating and validating surveys of character, and developing practical tools for individuals and practitioners.

Chapter 4: Mentor and Provider

AccessScience accessscience.com

AccessScience is an authoritative and dynamic online resource that contains incisively written, high-quality reference material that covers all major scientific disciplines. An award-winning gateway to scientific knowledge, it offers links to primary research material, videos and exclusive animations, plus specially designed curriculum maps for teachers. With these and other online features, AccessScience is continually expanding the ways it can demonstrate and explain core, trustworthy scientific information in a way that inspires and guides users to deeper knowledge.

Atlas Obscura atlasobscura.com

At Atlas Obscura, our mission is to inspire wonder and curiosity about the incredible world we all share.

Biodiversity Information System for Europe biodiversity.europa.eu

BISE is a single entry point for data and information on biodiversity supporting the implementation of the EU strategy and the Aichi targets in Europe.

Biomimicry Institute biomimicry.org

Our Mission: The purpose of the Biomimicry Institute is to naturalize biomimicry in the culture by promoting the transfer of ideas, designs, and strategies from biology to sustainable human systems design.

Center for Biologically Inspired Design cbid.gatech.edu

CBID is an interdisciplinary center for research and development of design solutions that occur in biological processes. Founded in 2005, it is one of more than 100 interdisciplinary research units funded at Georgia Institute of Technology.

Cicada Mania cicadamania.com

Dedicated to cicadas, the most amazing insects in the world.

Do Lectures thedolectures.com

The idea is a simple one. That people who Do things can inspire the rest of us to go and Do things too. So each year, we invite a set of people to come and tell us what they Do.

Earth Island Institute earthisland.org

Earth Island Institute is a non-profit, public interest, membership organization that supports people who are creating solutions to protect our shared planet.

Ecology and Society ecologyandsociety.org

Ecology and Society is an electronic, peer-reviewed, multi-disciplinary journal devoted to the rapid dissemination of current research. Manuscript submission, peer review, and publication are all handled on the Internet. Software developed for the journal automates all clerical steps during peer review, facilitates a double-blind peer review process, and allows authors and editors to follow the progress of peer review on the Internet.

Environment and Ecology environment-ecology.com

Sections: Environment, Ecology, Nature, Geography, Habitat, Gaia & Holistic View, Systems Theory, Energy & Economy, Sustainable Development, World Heritage.

Fountain Magazine fountainmagazine.com

Published bimonthly and distributed throughout the world, *The Fountain* covers themes on life, belief, knowledge, and universe. In an age of overspecialization in learning and over-indulgence in day-to-day occupations, *The Fountain*'s discourse refers to an overarching coverage of the human life with content as diverse and rich as the human life itself, yet with a common thread and pattern that is neatly knitted all the way through our diverse departments under humanities and sciences.

GreenBiz greenbiz.com

GreenBiz advances the opportunities at the intersection of business, technology and sustainability. Through its websites, events, peer-to-peer network and research, GreenBiz promotes the potential to drive transformation and accelerate progress—within companies, industries and in the very nature of business.

Green Facts greenfacts.org/en/index.htm

Our Mission is to bring the factual content of complex scientific consensus reports on health and the environment to the reach of non-specialists.

Journal of Sustainability Education susted.com/wordpress/

The *Journal of Sustainability Education* (JSE) serves as a forum for academics and practitioners to share, critique, and promote research, practices, and initiatives that foster the integration of economic, ecological, and social-cultural dimensions of sustainability within formal and non-formal educational contexts.

Millennium Ecosystem Assessment millenniumassessment.org/en/index.html

The Millennium Ecosystem Assessment assessed the consequences of ecosystem change for human well-being. From 2001 to 2005, the MA involved the work of more than 1,360 experts worldwide. Their findings provide a state-of-the-art scientific appraisal of the condition and trends in the world's ecosystems and the services they provide, as well as the scientific basis for action to conserve and use them sustainably.

Nature nature.com

Aim and Scope: *Nature* is a weekly international journal publishing the finest peer-reviewed research in all fields of science and technology on the basis of its originality, importance, interdisciplinary interest, timeliness, accessibility, elegance and surprising conclusions. *Nature* also provides rapid, authoritative, insightful and arresting news and interpretation of topical and coming trends affecting science, scientists and the wider public.

Nature Mentoring nature-mentor.com

When I was 15 years old I had an experience of sudden lucid clarity while hiking in the woods. Since then I've been passionately seeking tools for helping modern humans develop razor sharp natural instincts. I'm the author of multiple courses & ebooks about bird language, naturalist training, observation skills & outdoor mindfulness. My goal is to share these life changing skills with YOU!

James A. Swan's Home Page jamesswan.com

As a writer, producer and performer, James has sought to use mass media to promote conservation of natural resources, support law enforcement, educate about health, and provide entertainment to nourish the soul.

Twisted Sifter twistedsifter.com

The Sifter has one objective: To educate, entertain, and inspire each and every day.

Chapter 5: Nature's Intelligence

Audubon audubon.org

The National Audubon Society protects birds and the places they need, today and tomorrow, throughout the Americas using science, advocacy, education, and on-the-ground conservation.

Crazy About Mushrooms blog.crazyaboutmushrooms.com

Anna McHugh is a radio journalist and mycophile (mushroom lover) from Nevada City, CA. While living in the Northwest, Anna discovered a passion for mycology, and started hunting wild mushrooms, growing gourmet mushrooms at home, and learning botany.

Kosmos kosmosjournal.org

Our mission at Kosmos Associates is to inform, inspire and engage individual and collective participation for global transformation in harmony with all Life. We do this by sharing transformational thinking and policy initiatives, aesthetic beauty and collective wisdom.

Mathematical Association of America maa.org

The Mathematical Association of America is the world's largest community of mathematicians, students, and enthusiasts. We further the understanding of our world through mathematics because mathematics drives society and shapes our lives. The mission of the MAA is to advance the understanding of mathematics, and its impact on the world.

Mongabay news.mongabay.com

Mongabay is an environmental science and conservation news and information site. Much of Mongabay has operated under a non-profit—Mongabay.org—since 2012.

Plant Wisdom plantwisdom.org

We believe: (1) Plants are wise. (2) We are wise when we study plants. (3) We are wise when we pass this interest to our children.

PLOS ONE journals.plos.org

The world's first multidisciplinary Open Access journal, *PLOS ONE* accepts scientifically rigorous research, regardless of novelty. *PLOS ONE*'s broad scope provides a platform to publish primary research, including interdisciplinary and replication studies as well as negative results. The journal's publication criteria are based on high ethical standards and the rigor of the methodology and conclusions reported. Scope: *PLOS ONE* features reports of original research from the natural sciences, medical research, engineering, as well as the related social sciences and humanities that will contribute to the base of scientific knowledge. By not excluding research on the basis of subject area, *PLOS ONE* facilitates the discovery of connections...whether within or between disciplines.

Tree Whispering treewhispering.com

Mission: Inspire people to appreciate trees and plants as living Beings and to take respectful actions toward them. Offer people a deeply personal—sometimes profound—experience of connection between their core selves and the living Beings of Nature. Change people's practices and transform their consciousness to come from tree or plant's point of view. Provide innovative ways to strengthen trees, forests, and all plants by healing their internal functionality.

Whale and Dolphin Conservation us.whales.org

WDC, Whale and Dolphin Conservation, is the leading charity dedicated to the protection of whales and dolphins. Our vision is a world where every whale and dolphin is safe and free. Our mission is to amaze people with the wonder of whales and dolphins and inspire global action to protect them.

Chapter 6: Kinship and Creativity

Biophilia Educational Project biophiliaeducational.org

The Biophilia Educational Project is a large-scale pilot project that builds on the participation of academics, scientists, artists, teachers and students at all academic levels. It is based around creativity as a teaching and research tool, where music, technology and the natural sciences are linked together in an innovative way.

Biophilia Foundation biophiliafoundation.org

The work of the Biophilia Foundation is premised on the belief that only private landowners' efforts to restore and protect natural resources, especially wildlife habitat, will recover the living resources of the degraded lands and watersheds of our country. While economically sound working lands are essential to our well being, so is the health and proper function of the natural systems upon which our economy and our existence are based.

Creativity Research Journal
explore.tandfonline.com/content/ed/hcrj-article-collection-2016

Aims and Scope: *Creativity Research Journal* publishes high-quality, scholarly research capturing the full range of approaches to the study of creativity—behavioral, clinical, cognitive, crosscultural, developmental, educational, genetic, organizational, psychoanalytic, psychometrics, and social.

EdenKeeper edenkeeper.org

Mission: EdenKeeper seeks to foster communication and connection between religious/spiritual people and the environment.

Fulfillment Daily fulfillmentdaily.com

Our Mission: To inspire you with tools for a fulfilling life through science-backed news you can trust. Our Vision: That you apply the knowledge you learn on our site to benefit your own life, inspire others, and, ultimately, contribute to uplifting all of society. We envision a world in which everyone has access to the science of fulfillment.

Good Nature Travel goodnature.nathab.com

Good Nature Travel is the official travel blog of Natural Habitat Adventures and World Wildlife Fund. NHA is a nature and adventure travel company that focuses on seeing the world's most amazing natural places with minimal impact. WWF is the world's leading environmental conservation organization.

Heleo heleo.com

At Heleo, we believe you start with people. Specifically, thought leaders—both established and up-and-coming—who engage, inspire, and motivate readers to make changes in their lives and the world. Neuroscientists, historians, biologists, economists, journalists, psychologists, entrepreneurs...Heleo thought leaders come from a wide range of backgrounds. What they all share, however, is a love of discovery, of investigation and exploration, and equally important, a love of communicating their stories and insights in entertaining, memorable ways.

Land Art Generator Initiative landartgenerator.org

The goal of the Land Art Generator is to accelerate the transition to post-carbon economies by providing models of renewable energy infrastructure that add value to public space, inspire, and educate—while providing equitable power to thousands of homes around the world.

Nature World News natureworldnews.com

Nature World News offers fascinating and comprehensive news about the scientific world. Whether it's about animals, health, space, or archaeological finds, the website brings out the science geek in every reader, fostering an improved appreciation of our environment.

One Green Planet onegreenplanet.org

One Green Planet is a platform for the growing compassionate and eco-conscious generation. Our goal is to redefine "green," because, well, the earth needs a media company that's really got its back! Our goal is to help create a world where we eat delicious food and use amazing products that provide us with maximum benefit and have minimum impact on the planet.

Open for Ideas openforideas.org

And that's what this site is about. It's about opening up creativity for everyone. It's about sharing practical skills and techniques so that we can all get better at coming up with ideas. It's about sharing the most sought-after business skill with the world. We aim to always be practical and fact-based. And promise never to tell you how to "seduce your muse."

Project Learning Tree plt.org

Mission: Project Learning Tree advances environmental literacy and promotes stewardship through excellence in environmental education, professional development, and curriculum resources that use trees and forests as windows on the world. Vision: Project Learning Tree is committed to creating a future where the next generation values the natural world and has the knowledge and skills necessary to make informed decisions and take responsible actions to sustain forests and the broader environment.

Terrapin Bright Green: 14 Patterns of Biophilic Design
terrapinbrightgreen.com/reports/14-patterns/
Abstract: Biophilic design can reduce stress, enhance creativity and clarity of thought, improve our well-being and expedite healing; as the world population continues to urbanize, these qualities are ever more important. Theorists, research scientists, and design practitioners have been working for decades to define aspects of nature that most impact our satisfaction with the built environment. "14 Patterns of Biophilic Design" articulates the relationships between nature, human biology and the design of the built environment so that we may experience the human benefits of biophilia in our design applications.

THNK thnk.org
THNK is a School of Practice and a School of Thought. THNK designs and facilitates transformational in-person learning experiences to train and support global intrapreneurs and entrepreneurs in developing the mindsets, skillsets, and toolsets needed to scale their impact on the world's biggest challenges.

Thrive thrivemovement.com
The mission of Clear Compass Media, LLC is to create compelling media that assists humanity in thriving through the evolutionary challenges threatening our survival. This website and the documentary film THRIVE: What on Earth Will It Take? are the main projects of Clear Compass Media. We are a small team of committed individuals based out of Soquel, CA, dedicated to getting the message of THRIVE out to as many people worldwide as possible.

Thrive Global journal.thriveglobal.com
More than living. Thriving.

Tree of Life Web Project tolweb.org
The Tree of Life Web Project is a collection of information about biodiversity compiled collaboratively by hundreds of expert and amateur contributors. Its goal is to contain a page with pictures, text, and other information for every species and for each group of organisms, living or extinct. Connections be-

tween Tree of Life web pages follow phylogenetic branching patterns between groups of organisms so visitors can browse the hierarchy of life and learn about phylogeny and evolution as well as the characteristics of individual groups.

The Second Principle—The Work of Leslie Owen Wilson, EdD
thesecondprinciple.com
I bid you warm greetings and invite you to view my personal cache of informational educational webpages and blog space. Here are posted updated webpages from my original site, which was housed at the University of Wisconsin-Stevens Point from 1997–2013. Many of the pages contained herein are newer, updated versions of concepts and ideas that were used to support my teaching of many varied education courses during my 22+ year career as a university professor.

Chapter 7: Compassion and Coexistence

Adore Animals adoreanimals.com
Adore Animals is a volunteer, not-for-profit website which aims to inform and educate about animal issues and inspire positive solutions for change through information. Adore Animals promotes a culture that protects, respects and cares for all animals. We encourage animal organisations, corporates and communities to work together to find suitable strategies to help animals in need, while fostering positive relationships between humans and animals across all ages and specific needs groups.

Marc Bekoff, PhD marcbekoff.com
Marc Bekoff is professor emeritus of Ecology and Evolutionary Biology at the University of Colorado, Boulder, and is a Fellow of the Animal Behavior Society and a past Guggenheim Fellow. In 2000 he was awarded the Exemplar Award from the Animal Behavior Society for major long-term contributions to the field of animal behavior. Marc is also an ambassador for Jane Goodall's Roots & Shoots program, in which he works with students of all ages, senior citizens, and prisoners, and also is a member of the Ethics Committee of the Jane Goodall Institute.

BioScience academic.oup.com/bioscience
Since 1964, *BioScience* has presented readers with timely and authoritative overviews of current research in biology, accompanied by essays and discussion sections on education, public policy, history, and the conceptual underpinnings of the biological sciences.

Care2 care2.com

Care2 was founded with a simple mission: to help make the world a better place. Based in Silicon Valley, we are the world's largest social network for good. We are a community of over 40 million standing together, starting petitions and sharing stories that inspire action. For the past 19 years, Care2 has been a pioneer in online advocacy and we continue to focus on being on the forefront of creating technology that connects people to ways to make a difference in the world.

Center for Compassion and Altruism Research and Education
ccare.stanford.edu

Mission: CCARE investigates methods for cultivating compassion and promoting altruism within individuals and society through rigorous research, scientific collaborations, and academic conferences. In addition, CCARE provides a compassion cultivation program and teacher training as well as educational public events and programs.

Center for Compassion Studies compassioncenter.arizona.edu

Our founding Director, Charles Raison, MD, a psychiatrist in the UA College of Medicine, and renowned research scientist of mind-body approaches to mental health, has long been deeply interested in the modern application of ancient healing and spiritual traditions to alleviate mental and physical suffering.... Through his vision, the Center for Compassion Studies was created to expand the availability of compassion training methods, and to further investigate our understanding of the impact of these powerful methods on the wellbeing of humanity and our environment.

Charter for Compassion charterforcompassion.org

Our organization—Charter for Compassion—was inspired by the Charter for Compassion, created by Karen Armstrong and the Council of Conscience in 2009, and inherits a confluence of contributions made by TED.com, the Compassionate Action Network, the Fetzer Institute, and many others. Charter for Compassion provides an umbrella for people to engage in collaborative partnerships worldwide. Our mission is to bring to life the principles articulated in the Charter for Compassion through concrete, practical action in a myriad of sectors.

Compassionate Conservation compassionateconservation.org

This site offers a window on the emerging field of Compassionate Conservation, which aims to promote the consideration of animal welfare in a conservation context.

Conservation India conservationindia.org

CI is a non-profit, non-commercial portal that aims to facilitate nature conservation by providing reliable information, and the tools needed to campaign effectively. We define conservation as knowledge-driven actions that lead to the effective management and recovery of wildlife. That means giving priority to meeting the ecological needs of wildlife populations in decline, and to the recovery and expansion of their habitats. CI is committed to promoting conservation strategies that are rooted in evidence.

The Dalai Lama Center for Peace + Education dalailamacenter.org

During the Dalai Lama's visit to Vancouver, Canada, in 2004, we began a dialogue about the future of education in the 21st Century. The Dalai Lama stated his unequivocal belief that formal education must balance educating the mind with educating the heart. By doing so, we can foster the conditions for a more compassionate and peaceful future. Following the 2004 visit, His Holiness endorsed the founding of the The Dalai Lama Center for Peace and Education (DLC) in Vancouver. He specifically requested that the Center gather scientific research and develop evidence-informed programs to show the world that educating the heart is beneficial to both the individual and the communities we live in.

Office of His Holiness The Dalai Lama dalailama.com

The Dalai Lamas are believed by Tibetan Buddhists to be manifestations of Avalokiteshvara or Chenrezig, the Bodhisattva of Compassion and the patron saint of Tibet. Bodhisattvas are realized beings, inspired by the wish to attair complete enlightenment, who have vowed to be reborn in the world to help all living beings.

European Green Belt europeangreenbelt.org

The European Green Belt, our shared natural heritage along the line of the former Iron Curtain, is to be conserved and restored as an ecological network connecting high-value natural and cultural landscapes while respecting the economic, social and cultural needs of local communities.

Evolver evolver.net

Evolver is a platform for content, learning, and commerce serving a global community of conscious consumers seeking optimal states of well being in mind, body, and spirit. We serve as a trust-agent for individuals and groups seeking to connect in this transformative culture, one of wisdom, beauty, and fun.

Garrison Institute garrisoninstitute.org

The mission of the Garrison Institute is to demonstrate and disseminate the importance of contemplative practices and spiritually grounded values in building sustainable movements for a healthier, safer, and more compassionate world. Working collaboratively with practitioners in diverse fields, the Institute develops and hosts retreats and symposia, produces research and publications, and provides a hub for ongoing learning networks.

Good goodinc.com

GOOD is a global media brand and social impact company. Our collective mission is to help people and organizations be forces for good. Through award-winning media and creative partnerships, we connect deeply and authentically with this generation's desire for purpose.

TheHumanist.com thehumanist.com

TheHumanist.com is the daily online news site of the American Humanist Association. Founded in 2014 upon the merger of the American Humanist Association's weekly e-zine Humanist Network News and the website of the *Humanist* magazine, TheHumanist.com is the online hub for news, politics, science, and culture from a humanist perspective.

Live Science livescience.com

For the science geek in everyone, Live Science offers a fascinating window into the natural and technological world, delivering comprehensive and compelling news and analysis on everything from dinosaur discoveries, archaeological finds and amazing animals to health, innovation and wearable technology.

Mind & Life Institute mindandlife.org

The mission of the Mind & Life Institute is to alleviate suffering and promote flourishing by integrating science with contemplative practice and wisdom traditions.

The Natural Contemplative blog.naturalcontemplative.com

I have spent my life asking questions about our place on Earth; working for peace, social justice and environmental conservation; living a contemplative, listening life; and sharing my experiences through writing and education. I have studied the vocal behavior of marine mammals and songbirds. I now devote much of my time to recording natural soundscapes and composing and performing music that is inspired by those soundscapes.

Nature Conservation Foundation ncf-india.org

At the Nature Conservation Foundation, our goal is to contribute to the knowledge and conservation of India's unique wildlife heritage with innovative research and imaginative solutions. We work in a range of wildlife habitats, from coral reefs and tropical rainforests to the high mountains of the Himalaya.

Nonhuman Rights Project nonhumanrights.org

Mission: The Nonhuman Rights Project is the only civil rights organization in the United States working through litigation, public policy advocacy, and education to secure legally recognized fundamental rights for nonhuman animals.

Omega Institute eomega.org

Mission: Through innovative educational experiences that awaken the best in the human spirit, Omega provides hope and healing for individuals and society.

OnEarth archive.onearth.org

OnEarth (pronounced "on earth," not "one earth") is a quarterly magazine and daily online publication of thought and opinion about the environment. It explores the contemporary environmental landscape through the lenses of science, public health, technology, culture, business, food, and politics. Our original, in-depth reporting examines the challenges that confront our earth and its inhabitants, and our expert analysis evaluates the most promising solutions.

PrimeMind primemind.com

At PrimeMind we consider the questioning of culture, behavior, and surroundings vital to the human experience. We strive to deliver ideas that provoke thought and feeling.

Project Coyote projectcoyote.org

Project Coyote is a national non-profit organization based in Northern California whose mission is to promote compassionate conservation and coexistence between people and wildlife through education, science and advocacy.

The Rewilding Institute rewilding.org/rewildit/

The Rewilding Institute Website is the essential source of information about the integration of traditional wildlife and wildlands conservation with conservation biology to advance landscape-scale conservation. It provides explanations of key concepts with downloadable documents and links to important papers, essential books, and many groups working on various continental-scale conservation initiatives in North America.

The Safina Center safinacenter.org

At the Safina Center we advance the case for life on Earth by fusing scientific understanding, emotional connection, and a moral call to action. We create an original blend of science, art, and literature in the form of books and articles, scientific research, photography, films, sound-art, and spoken words. We bear first-hand witness and then we speak up, we speak out, and we teach. Our work is designed to inspire and engage others to devote their time and energies to conservation of wild things and wild places.

Think Kindness thinkkindness.org

Think Kindness inspires measurable acts of Kindness in schools and communities around the world. We believe that each act of Kindness, no matter how small, has an unforeseeable ripple effect that makes the world a better place. Every person on this planet can think of at least one Kind act that made your day a little brighter. It is our mission to inspire these Kind acts. To gather thousands of people to perform seemingly simple acts of Kindness in solidarity and that will result in a wave of good that will make a difference in the world.

United Nations Statistics Division unstats.un.org/home

The United Nations Statistics Division is committed to the advancement of the global statistical system. We compile and disseminate global statistical information, develop standards and norms for statistical activities, and support countries' efforts to strengthen their national statistical systems. We facilitate the coordination of international statistical activities and support the functioning of the United Nations Statistical Commission as the apex entity of the global statistical system.

WildEarth Guardians wildearthguardians.org

WildEarth Guardians protects and restores the wildlife, wild places, wild rivers, and health of the American West.

Chapter 8: An Ecocentric Ethic

The Life and Legacy of Rachel Carson rachelcarson.org

This website was created in 2000 to introduce the life and work of Rachel Carson (1907–1964), one of the 20th century's most important voices in literature and the environment. Few of the students whom I was teaching at the time had ever heard of Carson or of her revolutionary text, *Silent Spring*. I wanted to change that.

Earthbound People earthboundpeople.com

Who are the Earthbound people? The short answer is they are imaginary. The Earthbound people, or the people of Gaia are theoretical ideas proposed by French sociologist and science studies scholar Bruno Latour as part of his 2013 Gifford Lectures, "Facing Gaia: Six Lectures on the Political Theology of Nature." In essence, the Earthbound people are a way to help us think about a new political community that could emerge, which is rooted in...humans embedded in nature, rather than outside or separate from it.

HeartMath Institute heartmath.org

The mission of the HeartMath Institute is to help people bring their physical, mental and emotional systems into balanced alignment with their heart's intuitive guidance. This unfolds the path for becoming heart-empowered individuals who choose the way of love, which they demonstrate through compassionate care for the well-being of themselves, others and Planet Earth.

Sandra Ingerman sandraingerman.com

Sandra Ingerman, MA, is an award winning author of ten books, including *Soul Retrieval: Mending the Fragmented Self, Medicine for the Earth: How to Heal Personal and Environmental Toxins, Walking in Light: The Everyday Empowerment of Shamanic Life* and *Speaking with Nature* (co-authored with Llyn Roberts). She is the presenter of seven audio programs produced by Sounds True, and she is the creator of the Transmutation App.

Insight Meditation Center insightmeditationcenter.org

The Insight Meditation Center is a community-based, urban refuge for the teachings and practice of insight (mindfulness, vipassana) meditation. We offer Buddhist teachings in clear, accessible and open-handed ways.

International Society of Environmental Ethics enviroethics.org

This is the official site of the International Society for Environmental Ethics. For more than 20 years ISEE has striven to advance research and education in the field of environmental ethics and philosophy, and to promote appropriate human use, respect, conservation, preservation, and understanding of the natural world.

Wallace J. Nichols wallacejnichols.org

Dr. Wallace "J" Nichols, called "Keeper of the Sea" by *GQ Magazine* and "a visionary" by *Outside Magazine*, is an innovative, silo-busting, entrepreneurial scientist, movement maker, renowned marine biologist, voracious Earth and idea explorer, wild water advocate, bestselling author, sought after lecturer, and fun-loving Dad. He also likes turtles (a lot).

Resilience resilience.org

Resilience.org aims to support building community resilience in a world of multiple emerging challenges: the decline of cheap energy, the depletion of critical resources like water, complex environmental crises like climate change and biodiversity loss, and the social and economic issues which are linked to these. We like to think of the site as a community library with space to read and think, but also as a vibrant café in which to meet people, discuss ideas and projects, and pick up and share tips on how to build the resilience of your community, your household, or yourself.

Spirit Rock Meditation Center spiritrock.org

Spirit Rock Meditation Center is a spiritual education and training institution whose purpose is to: (1) Bring people to a depth of realization of the Buddha's path of liberation through direct experience. (2) Provide the community of practitioners with inspiration and teachings to integrate and manifest wisdom and compassion in all aspects of their lives, for the benefit of all beings.

UPLIFT upliftconnect.com

UPLIFT stands for Unity, Peace and Love In a Field of Transcendence.

Bibliography

Ackerman, Diane. *The Human Age: The World Shaped By Us*. W. W. Norton, 2015.

Ackerman, Jennifer. *The Genius of Birds*. Penguin Books, 2017.

Arvay, Clemens G. *The Biophilia Effect: A Scientific and Spiritual Exploration of the Healing Bond Between Humans and Nature*. Sounds True, 2018.

Barber, Nigel. *Kindness in a Cruel World: The Evolution of Altruism*. Prometheus Books, 2004.

Bateson, Gregory. *Mind and Nature: A Necessary Unity*. Hampton Press, 2002.

Bekoff, Marc. *The Animal Manifesto: Six Reasons for Expanding Our Compassion Footprint*. New World Library, 2010.

Bekoff, Marc, ed. *Ignoring Nature No More: The Case for Compassionate Conservation*. University of Chicago Press, 2013.

Bekoff, Marc. *Rewilding Our Hearts: Building Pathways of Compassion and Co-existence*. New World Library, 2014.

Bekoff, Marc and Jessica Pierce. *The Animals' Agenda: Freedom, Compassion, and Coexistence in the Human Age*. Beacon Press, 2017.

Benyus, Janine M. *Biomimicry: Innovation Inspired by Nature*. William Morrow, 1997.

Berry, Thomas. *The Sacred Universe: Earth, Spirituality, and Religion in the Twenty-First Century*. Columbia University Press, 2009.

Berry, Thomas. *Evening Thoughts: Reflecting on Earth as a Sacred Community*. Counterpoint, 2015.

Block, Peter. *Stewardship: Choosing Service Over Self-Interest*. Berrett-Koehler Publishers, 2013.

Brown, Brené. *Braving the Wilderness: The Quest for True Belonging and the Courage to Stand Alone*. Random House, 2017.

Butler, Tom and Sandra Lubarsky. *On Beauty: Douglas R. Tompkins—Aesthetics and Activism*. David Brower Center, 2017.

Buzzell, Linda and Craig Chalquist, eds. *Ecotherapy: Healing with Nature in Mind*. Counterpoint, 2009.

Capra, Fritjof. *The Turning Point: Science, Society, and the Rising Culture*. Bantam, 1984.

Capra, Fritjof. *The Web of Life: A New Scientific Understanding of Living Systems*. Anchor Books, 1997.

Capra, Fritjof and Luigi Luisi. *The Systems View of Life: A Unifying Vision*. Cambridge University Press, 2014.

Carson, Rachel L. *Silent Spring*. Mariner Books, 50th anniversary edition, 2002.

Carson, Rachel L. *The Sense of Wonder: A Celebration of Nature for Parents and Children*. Harper Perennial, 2017.

Chamovitz, Daniel. *What a Plant Knows: A Field Guide to the Senses*. Scientific American/Farrar, Straus and Giroux, 2012.

Christie, Douglas E. *The Blue Sapphire of the Mind: Notes for a Contemplative Ecology*. Oxford University Press, 2012.

Clayton, Susan and Gene Myers. *Conservation Psychology: Understanding and Promoting Human Care for Nature*. Wiley-Blackwell, 2015.

Clifford, M. Amos. *Your Guide to Forest Bathing: Experience the Healing Power of Nature*. Conari Press, 2018.

Clinebell, Howard. *Ecotherapy: Healing Ourselves, Healing the Earth*. Routledge, 1996.

Clingerman, Forrest, Brian Treanor, Martin Drenthen and David Utsler, eds. *Interpreting Nature: The Emerging Field of Environmental Hermeneutics*. Fordham University Press, 2013.

Coleman, Mark. *Awake in the Wild: Mindfulness in Nature as a Path of Self-Discovery*. New World Library, 2006.

Crist, Eileen and H. Bruce Rinker, eds. *Gaia in Turmoil: Climate Change, Biodepletion, and Earth Ethics in an Age of Crisis*. MIT Press, 2010.

Csikszentmihalyi, Mihaly. *Flow: The Psychology of Optimal Experience*. Harper Perennial, 2008.

Csikszentmihalyi, Mihaly. *Creativity: Flow and the Psychology of Discovery and Invention*. Harper Perennial, 2013.

Dalai Lama and Howard C. Cutler. *The Art of Happiness: A Handbook for Living*. Riverhead Books, 10th anniversary edition, 2009.

Darling, David. *The Universal Book of Mathematics: From Abracadabra to Zeno's Paradoxes*. John Wiley & Sons, 2004.

DeLuca, Denise Kelly. *Re-Aligning with Nature: Ecological Thinking for Radical Transformation*. White Cloud Press, 2016.

de Waal, Frans. *Are We Smart Enough to Know How Smart Animals Are?* W. W. Norton, 2017.

Elgin, Duane. *Awakening Earth: Exploring the Evolution of Human Culture and Consciousness*. William Morrow, 1993.

Emerson, Ralph Waldo. *Self-Reliance and Other Essays*. Dover Publications, unabridged edition, 1993.

Emerson, Ralph Waldo. *Nature and Selected Essays*. Penguin Classics, 2003.

Emerson, Ralph Waldo. *Nature*. CreateSpace Independent Publishing Platform, 2017.

Flores, Dan. *Coyote America: A Natural and Supernatural History*. Basic Books, 2017.

Frankl, Victor E. *Man's Search for Meaning*. Beacon Press, 2014.

Gardner, Gary T. *Inspiring Progress: Religion's Contributions to Sustainable Development*. W. W. Norton, 2006.

Gardner, Howard E. *Multiple Intelligences: New Horizons in Theory and Practice*. Basic Books, 2006.

Goleman, Richard, Lisa Bennett and Zenobia Barlow. *Ecoliterate: How Educators Are Cultivating Emotional, Social, and Ecological Intelligence*. Jossey-Bass, 2012.

Goodall, Jane. *Reason for Hope: A Spiritual Journey*. Grand Central Publishing, revised ed., 2000.

Hamma, Robert M. *Earth's Echo: Sacred Encounters with Nature*. Sorin Books, 2002.

Hawken, Paul. *The Ecology of Commerce: A Declaration of Sustainability*. HarperBusiness, revised edition, 2010.

Hawken, Paul, ed. *Drawdown: The Most Comprehensive Plan Ever Proposed to Reverse Global Warming*. Penguin Books, 2017.

Honoré, Carl. *In Praise of Slowness: Challenging the Cult of Speed*. HarperCollins, 2005.

Huppertz, Michael and Verena Schatanek. *Mindfulness in Nature*. CreateSpace Independent Publishing Platform, 2017.

Ingerman, Sandra and LLyn Roberts. *Speaking with Nature: Awakening to the Deep Wisdom of the Earth*. Bear & Company, 2015.

Jordan, Martin and Joe Hinds. *Ecotherapy: Theory, Research and Practice*. Palgrave Macmillan, 2016.

Kabat-Zinn, Jon. *Wherever You Go, There You Are: Mindfulness Meditation in Everyday Life*. Hachette Books, 10th anniversary edition, 2005.

Kahn, Peter H., Jr., and Patricia H. Hasbach, eds. *The Rediscovery of the Wild*. MIT Press, 2013.

Kellert, Stephen R. *Building for Life: Designing and Understanding the Human-Nature Connection*. Island Press, 2005.

Kellert, Stephen R. *Birthright: People and Nature in the Modern World*. Yale University Press, 2014.

Kellert, Stephen R. and Timothy J. Farham, eds. *The Good in Nature and Humanity: Connecting Science, Religion, and Spirituality with the Natural World*. Island Press, 2002.

Kellert, Stephen R., Judith H. Heerwagen and Martin L. Mador, eds. *Biophilic Design: The Theory, Science, and Practice of Bringing Buildings to Life*. John Wiley & Sons, 2008.

Kellert, Stephen R. and Edward O. Wilson. *The Biophilia Hypothesis*. Shearwater, 1995.

Keltner, Dacher. *Born to Be Good: The Science of a Meaningful Life*. W. W. Norton, 2009.

Keltner, Dacher, Jason Marsh and Jeremy Adam Smith, eds. *The Compassionate Instinct: The Science of Human Goodness*. W. W. Norton, 2010.

Kolbert, Elizabeth. *The Sixth Extinction: An Unnatural History*. Picador, 2015.

Kornfield, Jack. *No Time Like the Present: Finding Freedom, Love, and Joy Right Where You Are*. Atria Books, 2017.

Kuhn, Thomas S. *The Structure of Scientific Revolutions*. University of Chicago Press, 50th anniversary edition, 2012.

Lake-Thom, Bobby. *Spirits of the Earth: A Guide to Native American Nature Symbols, Stories, and Ceremonies*. Plume, 1997.

Lear, Linda. *Rachel Carson: Witness for Nature*. Mariner Books, 2009.

Leopold, Aldo. *A Sand County Almanac: With Essays on Conservation from Round River*. Ballentine Books, 1986.

Lewis, Charles A. *Green Nature/Human Nature: The Meaning of Plants in Our Lives*. University of Illinois Press, 1996.

Li, Qing. *Forest Bathing: How Trees Can Help You Find Health and Happiness*. Viking, 2018.

Lockhart, Kate. *Wildly Simple: Free Your Happiness Through the Power of Nature*. CreateSpace Independent Publishing Platform, 2017.

Louv, Richard. *Last Child in the Woods: Saving Our Children From Nature-Deficit Disorder*. Algonquin Books, updated and expanded edition, 2008.

Louv, Richard. *The Nature Principle: Reconnecting with Life in a Virtual Age*. Algonquin Books, 2012.

Louv, Richard. *Vitamin N: The Essential Guide to a Nature-Rich Life*. Algonquin Books, 2016.

Maturana, Humberto R. and Francisco J. Varela. *The Tree of Knowledge: The Biological Roots of Human Understanding*. Shambhala, revised edition, 1992.

McTaggart, Lynne. *The Intention Experiment: Using Your Thoughts to Change Your Life and the World*. Atria Books, 2008.

McTaggart, Lynne. *The Field: The Quest for the Secret Force of the Universe.* Harper Perennial, updated edition, 2008.
Meadows, Donella H. Diana Wright, ed. *Thinking in Systems: A Primer.* Chelsea Green Publishing, 2008.
Meine, Curt D. *Aldo Leopold: His Life and Work.* University of Wisconsin Press, 2010.
Miller, Peter. *The Smart Swarm: How to Work Efficiently, Communicate Effectively, and Make Better Decisions Using the Secrets of Flocks, Schools, and Colonies.* Avery, 2011.
Milton, John P. *Sky Above, Earth Below: Spiritual Practice in Nature.* Sentient Publications, 2006.
Miyazaki, Yoshifumi. *Shinrin Yoku: The Japanese Art of Forest Bathing.* Timber Press, 2018.
Narby, Jeremy. *Intelligence in Nature: An Inquiry into Knowledge.* Tarcher Perigee, 2006.
Neff, Kristin. *Self-Compassion: The Proven Power of Being Kind to Yourself.* William Morrow, 2015.
Nelson, Richard K. *Make Prayers to the Raven: A Koyukon View of the Northern Forest.* University of Chicago Press, 1986.
Nichols, Wallace J. *Blue Mind: The Surprising Science That Shows How Being Near, In, On, or Under Water Can Make You Happier, Healthier, More Connected, and Better at What You Do.* Back Bay Books, 2015.
Orr, David W. *Ecological Literacy: Education and the Transition to a Postmodern World.* State University of New York Press, 1991.
Orr, David W. *Earth in Mind: On Education, Environment, and the Human Prospect.* Island Press, 2004.
Orr, David W. *The Nature of Design: Ecology, Culture, and Human Intention.* Oxford University Press, 2004.
Patten, Terry. *A New Republic of the Heart: An Ethos for Revolutionaries—A Guide to Inner Work for Holistic Change.* North Atlantic Books, 2018.
Rosenzweig, Michael L. *Win-Win Ecology: How Earth's Species Can Survive in the Midst of Human Enterprise.* Oxford University Press, 2003.
Sabini, Meredith, ed. *The Earth Has a Soul: C. G. Jung on Nature, Technology & Modern Life.* North Atlantic Books, 2016.
Safina, Carl. *Beyond Words: What Animals Think and Feel.* Picador, 2016.
Selhub, Eva M. and Alan C. Logan. *Your Brain on Nature: The Science of Nature's Influence on Your Health, Happiness, and Vitality.* Collins, 2014.
Seppala, Emma. *The Happiness Track: How to Apply the Science of Happiness to Accelerate Your Success.* HarperOne, 2017.
Smiley, Nina and David Harp. *Mindfulness in Nature.* Hatherleigh Press, 2017.

Stamets, Paul. *Mycelium Running: How Mushrooms Can Help Save the World.* Ten Speed Press, 2005.

Starhawk. *The Earth Path: Grounding Your Spirit in the Rhythms of Nature.* HarperOne, 2005.

Stowe, John R. *The Findhorn Book of Connecting with Nature.* Findhorn Press, 2003.

Swan, James A. *Nature As Teacher and Healer: How to Reawaken Your Connection with Nature.* iUniverse, 2000.

Thompson, Mary Reynolds. *Reclaiming the Wild Soul: How Earth's Landscapes Restore Us to Wholeness.* White Cloud Press, 2014.

Thoreau, Henry David. *Walking.* Cricket House Books, 2010.

Thoreau, Henry David. *Walden and Civil Disobedience.* Signet, 2012.

Thorndike, Edward L. *Animal Intelligence: Experimental Studies.* Forgotten Books, 2017.

Tippett, Krista. *Becoming Wise: An Inquiry into the Mystery and Art of Living.* Penguin Press, 2017.

Tucker, Mary Evelyn and John Grim, eds. *Thomas Berry: Selected Writings on the Earth Community.* Orbis Books, 2014.

Uhl, Christopher. *Developing Ecological Consciousness: Paths to a Sustainable Future.* Rowman & Littlefield Publishers, 2003.

Varela, Francisco J., Evan Thompson and Eleanor Rosch. *The Embodied Mind: Cognitive Science and Human Experience.* MIT Press, revised edition, 2017.

Vaughan-Lee, Llewellyn, ed. *Spiritual Ecology: The Cry of the Earth.* The Golden Sufi Center, 2016.

von Kreisler, Kristin. *The Compassion of Animals: True Stories of Animal Courage and Kindness.* Three Rivers Press, 1999.

Walls, Laura Dassow. *Henry David Thoreau: A Life.* University of Chicago Press, 2017.

Weber, Andreas. *The Biology of Wonder: Aliveness, Feeling and the Metamorphosis of Science.* New Society Publishers, 2016.

Weber, Andreas. *Matter and Desire: An Erotic Ecology.* Chelsea Green Publishing, 2017.

White, Fred D., ed. *Essential Muir: A Selection of John Muir's Best Writings.* Heyday, 2006.

Williams, Florence. *The Nature Fix: Why Nature Makes Us Happier, Healthier, and More Creative.* W. W. Norton, 2017.

Wilson, Edward O. *Biophilia: The Human Bond with Other Species.* Harvard University Press, revised edition, 1986.

Wilson, Edward O. *The Meaning of Human Existence*. Liveright, 2015.
Wilson, Edward O. *The Origins of Creativity*. Liveright, 2017.
Wohlleben, Peter. *The Hidden Life of Trees: What They Feel, How They Communicate—Discoveries from a Secret World*. Greystone Books, 2016.
Wohlleben, Peter. *The Inner Life of Animals: Love, Grief, and Compassion—Surprising Observations of a Hidden World*. Greystone Books, 2017.

Notes

Foreword

1. James Loke Hale. "What Is The Google Doodle On Earth Day? Dr. Jane Goodall Contributed To The Stunning Image." Bustle, April 22, 2018. [Cited April 23, 2018] bustle.com/p/what-is-the-google-doodle-on-earth-day-dr-jane-goodall-contributed-to-the-stunning-image-8860956

Introduction

1. Stephen J. Gould. "Enchanted Evening." *Natural History*, September 1991, p. 14.
2. Rachel Carson. "Help Your Child to Wonder." *Woman's Home Companion*, July 1956, p. 46. [Cited April 6, 2018] training.fws.gov/History/Documents/carsonwonder.pdf
3. Stephen R. Kellert and Edward O. Wilson, eds. *The Biophilia Hypothesis*. Shearwater, 1995, p. 31.
4. Kevin Zelnio. "A World Ocean." *Scientific American*, June 8, 2011. [Cited April 7, 2018] blogs.scientificamerican.com/guest-blog/a-world-ocean/
5. Douglas E. Christie. *The Blue Sapphire of the Mind: Notes for a Contemplative Ecology*. Oxford University Press, 2012, p. xi.

Chapter 1

1. Wilma Mankiller. "Being Indigenous in the 21st Century." *Cultural Survival Quarterly*, 33:1, Spring 2009. [Cited March 2, 2018] culturalsurvival.org/publications/cultural-survival-quarterly/none/being-indigenous-21st-century
2. Melissa Kelly. "Teaching Students Who Have a Naturalist Intelligence." ThoughtCo, August 31, 2017. [Cited May 11, 2018] thoughtco.com/naturalist-intelligence-8098
3. Jack Kornfield. "Live in the Present." jackkornfield.com. [Cited May 15, 2018] jackkornfield.com/live-present/

4. Bernie Krause. "Notes on *Biophony*." Alonzo King Lines Ballet. [Cited May 4, 2018] linesballet.org/biophony
5. Thomas Berry. "Belonging: An Interview with Thomas Berry." *Parabola*, 24:1, February 1999, p. 26.
6. Chandra Taylor Smith. "Getting to the Soul of Our Connection to Nature." *The Huffington Post*, April 22, 2014, updated June 22, 2014. [Cited May 4, 2018] huffingtonpost.com/chandra-taylor-smith/getting-to-the-soul-of-ou_b_5188970.html
7. Ibid.
8. The Food Timeline. [Cited March 2, 2018] foodtimeline.org/index.html Also "Centers of Origin of Selected Crops." *Digital History*. [Cited May 4, 2018] digitalhistory.uh.edu/active_learning/explorations/columbus/columbian_answers_vegetables.cfm
9. "Discover John Muir." John Muir Trust. [Cited May 4, 2018] discoverjohnmuir.com
10. Emiliana R. Simon-Thomas. "Measuring Compassion in the Body." *Greater Good Magazine*, March 9, 2015. [Cited April 19, 2018] greatergood.berkeley.edu/article/item/measuring_compassion_in_the_body. Also Dacher Keltner. "The Compassionate Species." *Greater Good Magazine*, July 31, 2012. [Cited April 19, 2018] greatergood.berkeley.edu/article/item/the_compassionate_species
11. Jeremy Adam Smith. "Five Surprising Ways Oxytocin Shapes your Social Life." *Greater Good Magazine*, October 17, 2013. [Cited April 19, 2018] greatergood.berkeley.edu/article/item/five_ways_oxytocin_might_shape_your_social_life
12. Chris Mooney. "6 Amazing Ways Animals Show Compassion." *Mother Jones*, April 5, 2013. [Cited May 4, 2018] motherjones.com/environment/2013/04/6-amazing-ways-animals-show-compassion
13. Iain McGilchrist. "Meaning Is Not in Things, But in the Betweenness." *Resurgence and Ecologist*, 295, March/April 2016, p. 23. [Cited May 4, 2018] resurgence.org/magazine/issue295-walking-back-to-happiness.html

Chapter 2

1. Dictionary.com. [Cited May 4, 2018] dictionary.com/browse/awe?s=t
2. Caroline Gregorie. "How Awe-Inspiring Experiences Can Make You Happier, Less Stressed and More Creative." *The Huffington Post*, September 22, 2014. [Cited May 4, 2018] huffingtonpost.com/2014/09/22/the-psychology-of-awe_n_5799850.html. Also Dacher Keltner and

Jonathan Haidt. "Approaching awe, a moral, spiritual and aesthetic emotion." *Cognition and Emotion*, 17:2, 2003, pp. 297–314. [Cited May 4, 2018] greatergood.berkeley.edu/dacherkeltner/docs/keltner.haidt.awe.2003.pdf

3. Dacher Keltner. "Why Do We Feel Awe?" *Greater Good Magazine*, May 10, 2016. [Cited May 4, 2018] greatergood.berkeley.edu/article/item/why_do_we_feel_awe
4. Jason Silva. *Awe*. [Cited May 4, 2018] thisisjasonsilva.com Transcript at: *Shots of Awe*. Artemesia Shine. [Cited May 4, 2018] artemisiashine.com/shots-of-awe
5. Keltner. "Why Do We Feel Awe?"
6. *Wild Legacy: A Salute to the Remarkable Life of Douglas*. January 31, 2016. (5 min. 43 sec.–6 min. 11 sec.) [Cited May 4, 2018] tompkinsconservation.org/memorial/index.htm#tvideo
7. *Douglas Tompkins: A Wild Legacy*. [Cited March 1, 2018] tompkinsconservation.org/video/
8. Ralph Waldo Emerson. *Nature and Selected Essays*. Penguin, 2003, p. 47. Also "Beauty of Nature." *Nature*. Emerson Central. [Cited March 16, 2018] emersoncentral.com/texts/nature-addresses-lectures/nature2/beauty/
9. Emerson. *Nature and Other Essays*. Dover, 2009, p. 117. Also "Beauty." *The Conduct of Life*. Emerson Central. [Cited March 16, 2018] emersoncentral.com/texts/the-conduct-of-life/beauty/
10. Alfred Tauber. *Henry David Thoreau and the Moral Agency of Knowing*. University of California Press, 2003, pp. 75, 158.
11. Aldo Leopold. *A Sand County Almanac: With Essays on Conservation from Round River*. Ballantine Books, 1986, p. 262.
12. David Darling. *The Universal Book of Mathematics: From Abracadabra to Zeno's Paradoxes*. John Wiley & Sons, 2004, p. 34.

Chapter 3

1. National Wildlife Federation. *Connecting Kids and Nature: Health Benefits and Tips*. [Cited May 4, 2018] nwf.org/en/Kids-and-Family/Connecting-Kids-and-Nature/Health-Benefits-and-Tips
2. R.S. Ulrich. "View Through a Window May Influence Recovery from Surgery." *Science*, 224:4647, April 27, 1984, pp. 420–421. [Cited May 4, 2018] science.sciencemag.org/content/224/4647/420
3. Alexandra Sifferlin. "The Healing Power of Nature." *Time Magazine*, July 25, 2016. [Cited May 4, 2018] time.com/4405827/the-healing-power-of

-nature/. Summary by Denio Vale. Google +. [Cited May 4, 2018] plus.google.com/+DENIOVALE/posts/iq108,seqD5

4. Ibid.
5. Ibid.
6. Gregory N. Bratman, Gretchen C. Daily, Benjamin J. Levy, and James J. Gross. "The benefits of nature experience: Improved affect and cognition." *Landscape and Urban Planning*. 138, June 2015, pp. 41–50. [Cited May 4, 2018] sciencedirect.com/science/article/pii/S0169204615000286
7. Gretchen Reynolds. "How Walking in Nature Changes the Brain." *The New York Times*, July 22, 2015. [Cited May 4, 2018] well.blogs .nytimes.com/2015/07/22/how-nature-changes-the-brain/?_r=0. Also Gregory N. Bratman, J. Paul Hamilton, Kevin S. Hahn, Gretchen C. Daily and James J. Gross. "Nature Experience Reduces Rumination and Subgenual Prefrontal Cortex Activation." *Proceedings of the National Academy of Sciences of the United States of America*, July 2015, 112:28, pp. 8567–72. [Cited March 1, 2018] pnas.org/content/112/28/8567
8. Richard A. Fuller, Katherine N. Irvine, Patrick Devine-Wright, Philip H. Warren and Kevin J. Gaston. "Psychological Benefits of Greenspace Increase with Biodiversity." *Biology Letters*, 3:4, August 22, 2007. [Cited May 4, 2018] rsbl.royalsocietypublishing.org/content/3/4/390
9. *The Positive Effects of Nature on Well Being: Evolutionary Biophilia*. Positive Psychology Program, February 9, 2014. [Cited March 1, 2018] positivepsychologyprogram.com/why-nature-positively-affects-your -well-being-and-how-to-apply-it/
10. Kathleen Wolf and Katrina Flora. "Mental Health & Function." *Green Cities: Good Health*. College of the Environment, University of Washington, December 26, 2010, revised September 16, 2015. [Cited May 4, 2018] depts.washington.edu/hhwb/Thm_Mental.html
11. Richard Louv. *The Nature Principle: Reconnecting with Life in a Virtual Age*. Algonquin Books, 2012, pp. 46–7.
12. National Wildlife Federation. *Connecting Kids and Nature: Working Towards 10 Million Kids Outdoors*. [Cited June 7, 2017] nwf.org/What-We -Do/Kids-and-Nature.aspx and David Suzuki Foundation. *30x30 Nature Challenge. You + Nature*. [Cited May 4, 2018] 30x30.davidsuzuki.org
13. Erik Shonstrom. *What the Finns Know: "Friluftsliv" Gets Big Results in Finland's Schools*. Children & Nature Network, May 19, 2014. [Cited May 4, 2018] childrenandnature.org/2014/05/19/what-the-finns-know -friluftsliv-gets-big-results-in-finlands-schools/
14. Russel McLendon. *How 'Friluftsliv' Can Help You Reconnect with Nature.*

Mother Natural Network, June 27, 2014. [Cited March 1, 2018] mnn.com/earth-matters/wilderness-resources/blogs/how-friluftsliv-can-help-you-reconnect-with-nature

15. *Horticultural Therapy: History and Practice*. American Horticultural Therapy Association. [Cited May 4, 2018] ahta.org/horticultural-therapy
16. Joe Sempik, Rachel Hine and Deborah Wilcox, eds. *Green Care: A Conceptual Framework*. Loughborough University, April 2010, p. 38. [Cited May 4, 2018] agrarumweltpaedagogik.ac.at/cms/upload/bilder/green_care_a_conceptual_framework.pdf
17. Ibid., pp. 41–42.
18. Valerie F. Gladwell, Daniel K. Brown, Carly Wood, Gavin R. Sandercock and Jo L. Barton. "The Great Outdoors: How a Green Exercise Environment Can Benefit All." *Extreme Physiology & Medicine*, 2:3, January 3, 2013. [Cited March 1, 2018] ncbi.nlm.nih.gov/pmc/articles/PMC3710158/
19. Linda Buzzell and Craig Chalquist, eds. *Ecotherapy: Healing with Nature in Mind*. Counterpoint, 2009, p. 18.

Chapter 4

1. Merriam-Webster Dictionary. [Cited May 4, 2018] merriam-webster.com/dictionary/mentor
2. Janine M. Benyus. *Biomimicry: Innovation Inspired by Nature*. William Morrow, 1997.
3. "Technology that Imitates Nature." *The Economist: Technology Quarterly*, Q2, 2005, June 9, 2005. [Cited May 4, 2018] economist.com/node/4031083
4. Shea Gunther. "8 Amazing Examples of Biomimicry." *Mother Nature Network*, October 6, 2016. [Cited March 1, 2018] mnn.com/earth-matters/wilderness-resources/photos/7-amazing-examples-of-biomimicry/copying-mother-nature. Also Jessie Scanlon. "Janine Benyus's Theory of Evolution." *Bloomberg Business Week*, April 11, 2016. [Cited May 4, 2018] bloomberg.com/features/2016-design/a/janine-benyus/
5. Gunther. "8 Amazing Examples."
6. Merriam-Webster Dictionary. [Cited May 4, 2018] merriam-webster.com/dictionary/tipping%20point
7. Tom McKeag. "The Biomimicry Column: Is your system as resilient as nature's?" *GreenBiz*, August 6, 2013. [Cited March 1, 2018] greenbiz.com/blog/2013/08/06/your-system-resilient-natures. "Panarchy." Resilience Alliance. [Cited March 1, 2018] resalliance.org/panarchy

8. Diane Coutu. "How Resilience Works." *Harvard Business Review,* May 2002. [Cited May 4, 2018] hbr.org/2002/05/how-resilience-works
9. Victor E. Frankl. *Man's Search for Meaning.* Beacon Press, 2006, p. 112.
10. "Top 10 Oldest Animal Species on Earth." *The Mysterious World.* [Cited March 1, 2018] themysteriousworld.com/top-10-oldest-animal-species-on-earth/. and "Oldest Species on Earth is a Fern." *Phytophactor,* May 22, 2009. [Cited March 1, 2018] phytophactor.fieldofscience.com/2009/05/oldest-species-on-earth-is-fern.html
11. "Why Are Cicadas So Good at Math?" It's OK to Be Smart. PBS Digital Studio. [Cited May 4, 2018] youtube.com/watch?v=ivQaJwFRowc and Travis M. Andrews. "Billions of Cicadas Will Ascend Upon the Northeastern United States as Another 17-year Cycle Concludes." *Washington Post,* April 15, 2016. [Cited May 4, 2018] washingtonpost.com/news/morning-mix/wp/2016/04/15/billions-of-cicadas-will-descend-upon-the-northeastern-united-states-as-another-17-year-cycle-conclude/?utm_term=.47da7e098469. For more information, see *Cicada Mania.* [Cited March 1, 2018] cicadamania.com/where.html
12. "Adaptations of the Joshua Tree (*Yucca brevifolia*)." AccessScience, 2014. [Cited May 4, 2018] accessscience.com/content/adaptations-of-the-joshua-tree-yucca-brevifolia/BR0819145
13. Millennium Ecosystem Assessment. *Ecosystems and Human Well-being: Synthesis.* Island Press, 2005. [Cited March 1, 2018] millenniumassessment.org/documents/document.356.aspx.pdf
14. Ibid.
15. Paul Hawken. *The Ecology of Commerce: A Declaration of Sustainability.* HarperBusiness, 2010, p. 42–43.
16. Kenneth Brower. "A Voice for the Wilderness." *Earth Island Journal.* [Cited March 1, 2018] earthisland.org/journal/index.php/eij/article/wildness/

Chapter 5

1. David Abram. "In the Depths of a Breathing Planet" in Eileen Crist and H. Bruce Rinker, eds. *Gaia in Turmoil: Climate Change, Biodepletion and Earth Ethics in an Age of Crisis.* MIT Press, 2010, p. 225.
2. Jeremy Narby. *Intelligence in Nature: An Inquiry Into Knowledge.* Jeremy P. Tarcher, 2006, pp. 61, 85.
3. Robert Krulwich. "What Is It About Bees and Hexagons?" National Public Radio, May 14, 2013. [Cited May 4, 2018] npr.org/sections/krulwich/2013/05/13/183704091/what-is-it-about-bees-and-hexagons

and Frank Morgan. "Hales Proves Hexagonal Honeycomb Conjecture." Mathematical Association of America, June 17, 1999. [Cited May 4, 2018] maa.org/frank-morgans-math-chat-hales-proves-hexagonal-honeycomb-conjecture

4. Narby. *Intelligence in Nature*, pp. 55–67.
5. Ibid., pp. 112–121.
6. Ahmet D. Bahadir. "Optimization in Nature: Intelligent Solutions from Unintelligent Species." *The Fountain Magazine*, 94, July-August 2013. [Cited May 4, 2018] fountainmagazine.com/Issue/detail/optimization-in-nature-july-2013
7. Peter Miller. "The Genius of Swarms." *Archimorph*, June 3 2009. [Cited May 4, 2018] archimorph.com/2009/06/03/the-genius-of-swarms/
8. Ibid.
9. Kat McGowan. "Meet the Bird Brainiacs: American Crow." *Audubon Magazine*, March-April 2016. [Cited June 15, 2017] audubon.org/magazine/march-april-2016/meet-bird-brainiacs-american-crow
10. Michael Balter. "Meet the Brainiacs: Eurasian Jay." *Audubon Magazine*, March-April 2016. [Cited June 15, 2017] audubon.org/magazine/march-april-2016/meet-bird-brainiacs-eurasian-jay
11. Hannah Waters. "Can Pigeons Really Diagnose Cancer?" *Audubon*, November 23, 2015. [Cited June 15, 2017] audubon.org/news/can-pigeons-really-diagnose-cancer. Also Nicholas Bakalar. "Paging Dr. Pigeon: You're Needed in Radiology." *The New York Times*, November 24, 2015. [Cited May 4, 2018] nytimes.com/2015/11/25/science/pigeons-detect-breast-cancer-tumors.html?_r=0
12. Daniel Chamovitz. *What a Plant Knows: A Field Guide to the Senses*. Farrar, Straus and Giroux, 2012, pp. 137–8.
13. Narby. *Intelligence in Nature*, pp. 86–7. Also "What is brain plasticity, and can it help relieve psychiatric or degenerative brain disorders?" BrainFacts.org, April 23, 2012. [Cited March 2, 2018] brainfacts.org/ask-an-expert/what-is-brain-plasticity
14. Chamovitz. *What a Plant Knows*, pp. 116–119.
15. Ibid., pp.107–8.
16. Prudence Gibson. "Pavlov's Plants: New Study Shows Plants Can Learn from Experience." *The Conversation*, December 6, 2016. [Cited March 2, 2018] theconversation.com/pavlovs-plants-new-study-shows-plants-can-learn-from-experience-69794
17. Chamovitz. *What a Plant Knows*, p. 87.
18. Narby. *Intelligence in Nature*, pp. 95–7.

19. Anna McHugh. "Paul Stamets—The Intelligence of Mycelium." *Crazy About Mushrooms*. [Cited March 2, 2018] blog.crazyaboutmushrooms .com/paul-stamets-intelligence-mycelium/
20. Meredith Medland Sasseen. "Paul Stamets, Fungal Intelligence and the 21st Psychedelic Journey: How Mushrooms Can Help Save the World." *Living Green*. Personal Life Media. [Cited March 2, 2018] podcasts.personallifemedia.com/podcasts/224-living-green/episodes /2914-paul-stamets-fungal-intelligence-21st

Chapter 6

1. James A. Swan. "Nature As Teacher and Healer: How to Reawaken Your Connection to Nature." [Cited May 11, 2018] jamesswan.com /article-nature_as_teacher_and_healer.html Also James A. Swan. *Nature As Teacher and Healer: How to Reawaken Your Connection with Nature*. iUniverse, 2000, pp. xxiii–xxxii.
2. Erich Fromm. *The Anatomy of Human Destructiveness*. Holt, Rinehart and Winston, 1973, p. 366.
3. Stephen R. Kellert and Edward O. Wilson, eds. *The Biophilia Hypothesis*. Shearwater, 1995, p. 31.
4. Swan. *Nature As Teacher and Healer*.
5. Ibid.
6. Matt Stevens. "Video of Starving Polar Bear 'Rips Your Heart Out of Your Chest.'" *The New York Times*, December 11, 2017. [Cited February 5, 2018] nytimes.com/2017/12/11/world/canada/starving-polar -bear.html?action=click&contentCollection=Asia%20Pacific&module =Trending&version=Full®ion=Marginalia&pgtype=article
7. Stephen R. Kellert, Judith H. Heerwagen and Martin L. Mador, eds. Preface by Stephen R. Kellert and Judith H. Heerwagen. *Biophilic Design: The Theory, Science and Practice of Bringing Buildings to Life*. John Wiley & Sons, 2008, p. viii.
8. Bill Browning and Cary Cooper. "The Global Impact of Biophilic Design in the Workplace." *Architecture Now*, February 1, 2016. [Cited March 14, 2018] architecturenow.co.nz/articles/the-global-impact-of-biophilic -design-in-the-workplace/. Also "The Global Impact of Biophilic Design in the Workplace." Complete Report. [Cited March 14, 2018] interface .com/EU/en-GB/campaign/positive-spaces/human-spaces-report-en _GB
9. For information about the savanna hypothesis and biophilic design see Judith Heerwagen. "Psychosocial Value of Space." Whole Building

Design Guide, June 6, 2017. [Cited March 14, 2018] wbdg.org/resources/psychosocial-value-space

10. Browning and Cooper. "The Global Impact of Biophilic Design in the Workplace."
11. Jen Gadiel. "Greenbuild 2007 Quotes." *Green Design*, November 16, 2007. [Cited September 29, 2017] gadiel.com/greendesign/2007/11/greenbuild-2007-quotes.html
12. David Shenk. *Scott Barry Kaufman and David Shenk on the Messy Nature of Creativity*. Heleo Conversations, July 22, 2016. [Cited October 2, 2017] heleo.com/conversation-scott-barry-kaufman-and-david-shenk-on-the-messy-nature-of-creativity/10263/
13. Henry David Thoreau. "Walking." *The Atlantic*, June 1862. [Cited October 3, 2017] theatlantic.com/magazine/archive/1862/06/walking/304674/
14. Jill Suttie. "How Nature Can Make You Kinder, Happier, and More Creative." *Greater Good Magazine*, March 2, 2016. [Cited October 2, 2017] greatergood.berkeley.edu/article/item/how_nature_makes_you_kinder_happier_more_creative
15. "Nature Nurtures Creativity: Hikers More Inspired on Tests After Four Days Unplugged." U News Center, The University of Utah, December 12, 2012. [Cited October 2, 2017] archive.unews.utah.edu/news_releases/nature-nurtures-creativity-2/
16. Richard Coyne. "Soft Fascination." Reflections on Technology, Media and Culture. April 6, 2013. [Cited March 3, 2018] richardcoyne.com/2013/04/06/soft-fascination/. Also R. Kaplan and S. Kaplan. *The Experience of Nature: A Psychological Perspective*. Cambridge University Press, 1989.
17. Additional information: Michael Heizner. "Double Negative." [Cited October 6, 2017] doublenegative.tarasen.net/double-negative/. Christo and Jeanne-Claude. "Artworks: Realized Projects." [Cited October 18, 2017] christojeanneclaude.net/artworks/realized-projects
18. For information about land artists, see Carmen Zella. "Creating Connections with Nature Via Art." *Huffington Post: The Blog*, February 23, 2014; updated April 25, 2014. [Cited October 6, 2017] huffingtonpost.com/carmen-zella/creating-connections-withland-art_b_4833677.html
19. Martin Hill. "Art Practice." [Cited October 6, 2017] martin-hill.com/about/art-practice/
20. Jan Johnsen. "Willow Palaces, Cathedrals and Domes—Sanfte Strukturen." Serenity in the Garden, May 6, 2012. [Cited October 10, 2017]

serenityinthegarden.blogspot.com/2010/04/willow-palaces-cathedrals-and-domes.html. "Marcel Kalberer: Germany." Venus Architecture. [Cited October 10, 2017] venusarchitecture.com/en/marcel-kalberer.html. Also Sanfte Strukturen. sanftestrukturen.de. "Cattedrale Vegetale (Tree Cathedral)." Atlas Obscura. [Cited October 10, 2017] atlasobscura.com/places/cattedrale-vegetale-tree-cathedral

21. Land Art Generator Initiative. "LAGI's Founding." [Cited March 23, 2018] landartgenerator.org/founding.html
22. Land Art Generator Initiative. [Cited March 23, 2018] landartgenerator.org/index.html
23. Land Art Generator Initiative. "LAGI's Founding."

Chapter 7

1. Gray Cox. *Bearing Witness: Quaker Process and a Culture of Peace*. Pendle Hill Pamphlet 262. Pendle Hill Publications, 1985, 2014.
2. *The Sustainable Development Goals Report 2017*. United Nations Department of Economics and Social Affairs, Statistics Division, 2018. [Cited December 22, 2017] unstats.un.org/sdgs/report/2017/
3. Bjorn Carey. "Stanford Psychologists Show that Altruism is Not Simply Innate." *Stanford News*, December 18, 2014. [Cited December 22, 2017] news.stanford.edu/news/2014/december/altruism-triggers-innate-121814.html
4. "What is Altruism?" *Greater Good Magazine*, n.d. [Cited December 22, 2017] greatergood.berkeley.edu/altruism/definition
5. "Biological Altruism." *Stanford Encyclopedia of Philosophy*, June 3, 2003; substantive revision July 21, 2013. [Cited December 22, 2017] plato.stanford.edu/entries/altruism-biological/
6. Jonathan Birch. "Altruism in Nature." *The Philosophy of Social Evolution*. The Brains Blog, October 30, 2017. [Cited January 12, 2018] philosophyofbrains.com/2017/10/30/1-altruism-nature.aspx
7. Michael Greshko. "Why Female Vampire Bats Donate Blood to Friends." *National Geographic*, November 17, 2015. [Cited January 12, 2018] news.nationalgeographic.com/2015/11/151117-vampire-bats-blood-food-science-animals/
8. Sadie F. Dingfelder. "Altruism: An Accident of Nature?" American Psychological Association. *Monitor on Psychology*, 37(11), December 2006. Print version, page 44. [Cited January 12, 2018] apa.org/monitor/dec06/altruism.aspx
9. David G. Rand, Joshua D. Greene and Martin A. Nowak. "Spontaneous giving and calculated greed." *Nature*, 489, pp. 427–30, September 20,

2012. [Cited January 12, 2018] nature.com/articles/nature11467. Also Emiliana R. Simon-Thomas. "The Cooperative Instinct." *Greater Good Magazine*, September 21, 2012. [Cited January 12, 2018] greatergood .berkeley.edu/article/item/the_cooperative_instinct

10. Marc Bekoff. "Do Animals Have Emotions?" *The Bark*, November 2008. [Cited January 15, 2018] thebark.com/content/do-animals-have -emotions
11. Tia Ghose. "5 Animals with a Moral Compass." *LiveScience*, November 15, 2012. [Cited January 15, 2018] livescience.com/24800-animals-emotions -morality.html Also Virginia Morell. "It's Time to Accept That Elephants, Like Us, Are Empathetic Beings." *National Geographic*, February 23, 2014. [Cited January 15, 2018] news.nationalgeographic.com/news/2014/02/14 0221-elephants-poaching-empathy-grief-extinction-science/
12. Ghose. "5 Animals with a Moral Compass."
13. Marc Bekoff. "Animal Emotions: Exploring Passionate Natures." *Bio-Science*, 50:10, October 2000, p. 867. [Cited October 25, 2017] academic .oup.com/bioscience/article/50/10/861/233998/Animal-Emotions -Exploring-Passionate
14. Marc Bekoff. "The Nonhuman Rights Project: An Interview with Steven Wise." *Psychology Today*, December 16, 2016. [Cited January 15, 2018]. psychologytoday.com/blog/animal-emotions/201612/the-nonhuman -rights-project-interview-steven-wise
15. Ibid.
16. Philip Low et al. "The Cambridge Declaration on Consciousness." Francis Crick Memorial Conference on Consciousness in Human and Non-Human Animals, University of Cambridge, July 7, 2012. [Cited January 16, 2018] fcmconference.org/img/CambridgeDeclarationOn Consciousness.pdf
17. Marc Bekoff. "Compassionate Conservation Comes of Age." *Psychology Today*, November 15, 2017. [Cited January 16, 2018] psychologytoday .com/blog/animal-emotions/201711/compassionate-conservation -matures-and-comes-age
18. *Living with Elephants*. Evanescence Studios, July 24, 2014 (4 min. 20 sec.– 4 min. 27 sec.). [Cited January 17, 2018] youtube.com/watch?v=MWcdM jv41ho
19. Atula Gupta. "Giving Elephants the Space They Need, One SMS at a Time." *The Wire*, April 24, 2017. [Cited January 17, 2018] thewire.in /129142/elephants-bengal-deaths-kerala-drought/. Additional information at "The Elephant Hills." Nature Conservation Foundation. [Cited January 17, 2018] ncf-india.org/projects/in-the-elephant-hills

20. Ankita Sengupta. "Little to Fear from Leopards if not Provoked: Sanjay Gandhi National Park, Forest Officials." *Hindustan Times*, March 13, 2017. [Cited January 18, 2018] hindustantimes.com/mumbai-news/little-to-fear-from-leopards-if-not-provoked-say-sgnp-forest-officials/story-T5iCGVXpG35Zh8dnc5EkjJ.html
21. T.R. Shankar Raman. "Leopard Landscapes: Coexisting with Carnivores in Countryside and City." *Economic & Political Weekly*, 50:1, January 3, 2015. [Cited January 18, 2018] www.epw.in/journal/2015/1/reports-states-web-exclusives/leopard-landscapes.html-0
22. Ibid.
23. "Programs." Project Coyote. [Cited January 18, 2018] projectcoyote.org/programs/
24. Personal communication, March 2, 2018.
25. "Reforming Predator Management." Project Coyote. [Cited January 18, 2018] projectcoyote.org/programs/reforming-predator-management/
26. "Nonlethal Solutions to Reduce Conflicts." Project Coyote. [Cited January 18, 2018] projectcoyote.org/programs/ranching_with_wildlife/non lethal-solutions-reduce-conflicts/. Also "Coyote Friendly Communities." Project Coyote. [Cited January 18, 2018] projectcoyote.org/programs/coyote-friendly-communities/
27. Stacy Carlsen. Personal communication on March 22, 2018.
28. "Marin County Livestock & Wildlife Protection Program." Project Coyote. [Cited January 18, 2018] projectcoyote.org/project/marin-county-livestock-wildlife-protection-program/
29. "Coyote Friendly Communities." Project Coyote. [Cited January 18, 2018] projectcoyote.org/programs/coyote-friendly-communities/
30. Dave Foreman. "What Is Rewilding?" (adapted from Dave Foreman. *Rewilding North America*. Island Press, 2004). The Rewilding Institute. [Cited January 19, 2018] rewilding.org/rewildit/what-is-rewilding/
31. "Y2Y Achievements." Yellowstone to Yukon Conservation Initiative. [Cited January 19, 2018] y2y.net/vision/our-progress/y2y-achievements
32. "From Iron Curtain to Lifeline." European Green Belt. [Cited January 19, 2018] europeangreenbelt.org
33. "President of Chile and CEO of Tompkins Conservation Sign Decrees Creating 10 Million Acres of New National Parks." Tompkins Conservation. [Cited April 19, 2018] tompkinsconservation.org/news/2018/01/29/president-of-chile-and-ceo-of-tompkins-conservation-sign-decrees-creating-10-million-acres-of-new-national-parks/

34. Marc Bekoff. *Rewilding Our Hearts: Building Pathways of Compassion and Coexistence*. New World Library, 2014, p. 38.
35. Mary Reynolds Thompson. *Reclaiming the Wild Soul: How Earth's Landscapes Restore Us to Wholeness*. White Cloud Press, 2014, p. xxiii.

Chapter 8

1. Rachel Carson. "Help Your Child to Wonder." *Woman's Home Companion*, July 1956, p. 46. [Cited February 2, 2018] training.fws.gov/History/Documents/carsonwonder.pdf
2. His Holiness the 14th Dalai Lama of Tibet. "Universal Responsibility in the Modern World." (Transcript of His Holiness's Public Talk at Royal Albert Hall, London, United Kingdom, May 22, 2008). [Cited February 5, 2018] dalailama.com/messages/transcripts-and-interviews/universal-responsibility-modern-world
3. Fritjof Capra and Pier Luigi Luisi. *The Systems View of Life: A Unifying Vision*. Cambridge University Press, 2014, p. 65.
4. Kimberly Franklin. "An Introduction to Reconciliation Ecology." The Web of Life Project, 2007. [Cited February 5, 2018] tolweb.org/treehouses/?treehouse_id=4558
5. Carson, p. 46.
6. Selin Kesebir and Pelin Kesebir. "How Modern Life Became Disconnected from Nature." *Greater Good Magazine*, September 20, 2017. [Cited February 6, 2018] greatergood.berkeley.edu/article/item/how_modern_life_became_disconnected_from_nature
7. "Thousands petition junior dictionary over nature words." BBC News, December 21, 2017. [Cited February 6, 2018] bbc.com/news/uk-england-oxfordshire-42441025
8. Elizabeth Licata. "Erasing Nature." *Garden Rant*, October 17, 2017. [Cited February 6, 2018] gardenrant.com/2017/10/erasing-nature.html
9. Daniel Christian Wahl. "Learning from Nature and Designing as Nature: Regenerative Cultures Create Conditions Conducive to Life." The Biomimicry Institute, September 6, 2016. [Cited March 14, 2018] biomimicry.org/learning-nature-designing-nature-regenerative-cultures-create-conditions-conducive-life/
10. "Joseph Campbell." Pantheism Community, n.d. [Cited February 6, 2018] pantheism.com/about/luminaries/joseph-campbell/
11. Douglas E. Christie. *The Blue Sapphire of the Mind: Notes for a Contemplative Ecology*. Oxford University Press, 2012, p. xi.

12. "HeartMath Institute's Mission and Vision." HeartMath Institute. [Cited February 7, 2018] heartmath.org/about-us/hmi-mission/
13. Cherie Lindberg. "The case for heart-based living." Nature's Pathways, August 2013. [Cited March 14, 2017] naturespathways.com/editions/northeast-wi-edition/item/2662-the-case-for-heart-based-living#.Wqnxd2bMw_U
14. Lynne McTaggart. *The Field: The Quest for the Secret Force of the Universe*. Harper Perennial, 2008, p. xxiii.
15. Daniel A. Vallero. *Paradigms Lost: Learning from Environmental Mistakes, Mishaps and Misdeeds*. Butterworth-Heinemann, 2005, p. 387.
16. Starhawk. *The Earth Path: Grounding Your Spirit in the Rhythms of Nature*. HarperSanFrancisco, 2004, p. 5.
17. An interview with Wallace J. Nichols. Biophilic Cities. [Cited February 7, 2018] biophiliccities.org/blue-mind-an-interview-with-wallace-j-nichols/
18. Kevin Zelnio. "A World Ocean." *Scientific American*, June 8, 2011. [Cited February 15, 2018] blogs.scientificamerican.com/guest-blog/a-world-ocean/
19. Wallace J. Nichols. "Blue Marbles Project." [Cited February 7, 2018] wallacejnichols.org/130/blue-marbles-project.html
20. Bill Moyers. "Wendell Berry: Poet and Prophet." *Huffington Post*, October 2, 2013. [Cited February 9, 2018] huffingtonpost.com/bill-moyers/wendell-berry-poet--proph_b_4031836.html
21. Aldo Leopold. *A Sand County Almanac*. Ballantine Books, 1986, pp. 138–9. For information about Aldo Leopold's views on conservation see Curt D. Meine. *Aldo Leopold: His Life and Work*. University of Wisconsin Press, 2010.
22. "Captain Paul Watson—Saving Our Seas." Adore Animals. [Cited February 9, 2018] adoreanimals.com/captain-paul-watson-saving-our-seas/
23. David W. Orr. *Earth in Mind: On Education, Environment, and the Human Prospect*. Island Press, 2004, pp. 44–5.
24. Andrew Simms. "We Can Learn Resilience from the Natural World—But Only Up to a Point." *The Guardian*, September 3, 2012. [Cited February 12, 2018] theguardian.com/environment/2012/sep/03/100-months-resilience-natural-world

Index

About the Author

Andrés Edwards is an educator, award-winning author, media designer and sustainability consultant. He is founder and president of EduTracks, a firm specializing in designing and developing exhibit, print and education programs and offering consulting services on sustainable practices for nonprofit and business initiatives. He has worked as producer, exhibit developer and consultant for projects in natural history, biodiversity and sustainable community for companies and towns throughout the US and abroad.

He is the author of *The Heart of Sustainability: Restoring Ecological Balance from the Inside Out* (2015 Silver Award: Nautilus Book Awards), *Thriving Beyond Sustainability: Pathways to a Resilient Society* (2010 Gold Medal: Living Now Book Awards) and *The Sustainability Revolution: Portrait of a Paradigm Shift* (2005), which was selected by Apple to demonstrate the educational potential of ebooks used in conjunction with the iPad platform in academic settings. He is coauthor with Robert Z. Apte of *Tibet: Enduring Spirit, Exploited Land* (2004).

Andrés has given radio and television interviews and lectured and presented seminars about his work at conferences and universities and for business and community organizations. He lives in northern California.

For further information visit: andresedwards.com

A NOTE ABOUT THE PUBLISHER

New Society Publishers is an activist, solutions-oriented publisher focused on publishing books for a world of change. Our books offer tips, tools, and insights from leading experts in sustainable building, homesteading, climate change, environment, conscientious commerce, renewable energy, and more—positive solutions for troubled times.

DON'T EAT THIS BOOK (but you could)

We're proud to hold to the highest environmental and social standards of any publisher in North America. This is why some of our books might cost a little more. We think it's worth it!

- We print all our books in North America, never overseas
- All our books are printed on **100% post-consumer recycled paper**, processed chlorine-free, with low-VOC vegetable-based inks (since 2002)
- Our corporate structure is an innovative employee shareholder agreement, so we're one-third employee-owned (since 2015)
- We're carbon-neutral (since 2006)
- We're certified as a B Corporation (since 2016)

At New Society Publishers, we care deeply about *what* we publish—but also about *how* we do business.

Download our catalog at https://newsociety.com/Our-Catalog or for a printed copy please email info@newsocietypub.com or call 1-800-567-6772 ext 111.

New Society Publishers

ENVIRONMENTAL BENEFITS STATEMENT

For every 5,000 books printed, New Society saves the following resources:[1]

22	Trees
1,976	Pounds of Solid Waste
2,174	Gallons of Water
2,836	Kilowatt Hours of Electricity
3,592	Pounds of Greenhouse Gases
15	Pounds of HAPs, VOCs, and AOX Combined
5	Cubic Yards of Landfill Space

[1] Environmental benefits are calculated based on research done by the Environmental Defense Fund and other members of the Paper Task Force who study the environmental impacts of the paper industry.